W9-CHI-122

f **P**

COMM CHECK . . .

THE FINAL FLIGHT OF SHUTTLE COLUMBIA

Michael Cabbage *and* William Harwood

FREE PRESS
NEW YORK LONDON TORONTO SYDNEY

*f*P
FREE PRESS
A Division of Simon & Schuster, Inc.
1230 Avenue of the Americas
New York, NY 10020

All photographs unless otherwise noted are courtesy of NASA

For information about special discounts for bulk purchases,
please contact Simon & Schuster Special Sales:
1-800-456-6798 or business@simonandschuster.com

Designed by Joseph Rutt

Manufactured in the United States of America

10 9 8 7 6 5 4 3 2 1

Library of Congress Cataloging-in-Publication Data is available.

ISBN 0-7432-6091-0

To Mom and Dad
—Michael Cabbage

To the memory of Brian Welch
—William Harwood

ACKNOWLEDGMENTS

From Michael Cabbage: I would like to thank *Orlando Sentinel* publisher Kathy Waltz, editor Tim Franklin, and managing editor Elaine Kramer for making it possible for me to write this book and consistently supporting aerospace reporting as few others in American journalism; Bob Shaw, Alex Beasley, Sal Recchi and Sean Holton for their editing guidance, help and advice in the days and weeks following the disaster; fellow reporters and photographers on the *Sentinel's* Columbia coverage team, including Gwyneth Shaw, Robyn Suriano, Kevin Spear, Jim Leusner, Red Huber, Debbie Salamone, Dan Tracy, Jeff Kunerth, Scott Powers, Rene Stutzman, Pam Johnson, Mary Shanklin, Anthony Colarossi, Mike McLeod and Sean Mussenden, whose hard work helped uncover some of the events and issues described in this book; and special thanks to Sal Recchi and Ann Hellmuth for their friendship and moral support in the weeks following the accident. Thanks also to senior editor Craig Covault of *Aviation Week and Space Technology* magazine, another friend whose input was invaluable. And last but certainly not least, my thanks, love and eternal gratitude to my neglected wife, Lori, and children, Elizabeth and John, without whose love and support this book could not have been written.

From William Harwood: I would like to thank CBS News President Andrew Heyward, Al Ortiz and Mark Kramer for their long-standing commitment to covering the space program and for giving me time off to write this book. Special thanks also go to correspondent Sharyl Attkisson and her husband, Jim, for providing much-needed advice and moral support; and to Washington correspondent and aviation expert Bob Orr and producer Ward Sloane for going the extra mile to get the facts straight. Orr's perspective from long experience with past disasters—"complex systems fail in complex ways"—proved invaluable. Special thanks also to CBS Radio correspondent Peter King for his unfailing politeness and cheerful demeanor while holding the ravenous radio beast at bay; to Steven Young and Justin Ray of Spaceflightnow.com; and to Kathy Sawyer and Nils Bruzelius of *The Washington Post*. Finally, this project would not have been possible without the loving encouragement of my wife, Catherine, and our children, Houston and Riley. And of course, my parents, Kate and Walter Harwood, who taught me to wonder about the universe.

We would both like to thank our agent, Gail Ross, and our editor, Bruce Nichols, for their patience and encouragement. From NASA public affairs, thanks to Glenn Mahone and Bob Jacobs at NASA Headquarters, Johnson Space Center's James Hartsfield, Rob Navias and Kyle Herring, who arranged countless interviews, June Malone at the Marshall Space Flight Center, and Mike Rein and Bill Johnson at the Kennedy Space Center; Special thanks to Laura Brown of the Columbia Accident Investigation Board for providing technical feedback and setting up interviews; Kari Allen of Boeing; Mike Curry of United Space Alliance; Traci Lutter at Far Out Transcription Services; Phyllis Cabbage and Judy and David Mitchell for their Southern hospitality during six hectic weeks in Knoxville; the faculty of the University of Tennessee School of Journalism for their help and advice, past and present; and Charlie at The Tap Room, a friend to us and other wayward college students for more than a quarter-century.

CONTENTS

ONE: Re-Entry 1
TWO: Preparations 27
THREE: "Safe to Fly with No New Concerns" 53
FOUR: Launch 73
FIVE: A Shot in the Dark 93
SIX: Mixed Signals 125
SEVEN: Disaster 147
EIGHT: Aftermath 171
NINE: Echoes of Challenger 189
TEN: Re-Entry Revisited 225
ELEVEN: Returning to Flight 261
Epilogue 287
Appendices 299
Index 311

PREFACE

After seeing more than 40 shuttle landings, *Orlando Sentinel* space editor Michael Cabbage knew something was wrong.

Like other reporters at the Kennedy Space Center's shuttle runway, he had arrived an hour before Columbia's scheduled touchdown expecting a routine end to the 16-day flight. In fact, the handful of people who turned up at the landing strip that Saturday morning, Feb. 1, 2003, were more preoccupied with afternoon activities like yard work, sports events and cookouts than the return of the year's first shuttle mission.

But Cabbage had a sinking feeling. About 12 minutes remained before Columbia's 9:16 a.m. landing and Mission Control did not know where the shuttle was. Momentary "comm dropouts" were not unusual, but Columbia's crew had been out of contact with flight controllers in Houston for five long minutes. And that never had happened before.

As Cabbage watched a NASA television broadcast at the landing strip, he noticed something else peculiar: A red triangle used to show Columbia's progress across a map of North America had stopped moving over central Texas. That had to be a glitch, he thought. Moments later, his pager went off.

Cabbage ducked into a nearly-deserted public affairs center at the runway's mid-point to return the call. It was Robyn Suriano, another *Orlando Sentinel* reporter, who was monitoring the landing from the newspaper's bureau at Kennedy's press site.

"They can't contact the crew," Suriano said. "Should we be worried?"

"I know," Cabbage replied. "This isn't good."

As Cabbage spoke at a desk in the rear of the building, a voice suddenly boomed out of a nearby speakerphone dialed into an internal public affairs teleconference. He instantly recognized the voice of Kyle Herring, a public affairs officer at the Johnson Space Center in Houston.

"All of the sensors that dropped off before we lost contact were in the left wing," Herring said. "There are reports online of debris over Texas. We're getting ready to lock the doors and declare a contingency."

This can't be happening, Cabbage thought. Herring's terse comments meant flight controllers believed the shuttle and its crew were lost.

At the CBS News bureau a mile from the shuttle runway, William Harwood was putting the finishing touches on a post-mission wrap-up story he planned to post to the network's website within minutes of Columbia's touchdown. As always on landing day, Harwood was in the bureau's second-floor television studio, wired up and prepared to go on the air in the event of a disaster. But CBS had no plans to interrupt its Saturday morning news show for an uneventful landing and Harwood, a veteran of 107 of the 113 shuttle missions to date, did not expect any problems.

Space shuttle re-entries did not generate the same level of nervous anticipation as launches and Harwood was listening to Mission Control with, quite literally, half an ear. One of his two earpieces was plugged into a network audio loop in New York while the other was plugged into NASA's Mission Control circuit.

At 8:59 a.m., 17 minutes before landing, he heard Mission Control call Columbia about a landing gear tire pressure reading. He heard the commander begin to respond before a communications dropout cut off his reply. Harwood mentally logged the call and continued typing. But he was now listening for communications to be restored.

At 9:02 a.m., Justin Ray, a reporter with Spaceflightnow.com, sent Harwood an "Instant Messenger" note via the Internet:

"Should we be getting nervous?"

"I'm getting nervous," Harwood typed, wishing he could emphasize the "I."

Seconds later, Ray forwarded an Instant Messenger note from Stephen Clark, a Texas high school student watching Columbia streak across the sky southeast of Dallas. Harwood glanced at the note. Then he stared at it, his heart suddenly pounding.

CLARK: O M G [Oh My God] !!!!!!!!!!!!!!
CLARK: DUDE OMG
CLARK: It broke up!
CLARK: The shuttle broke up!

RAY: He's in Texas
HARWOOD: What!!!
HARWOOD: Are you serious?
RAY: Yes
RAY: North of Houston

It was a moment both reporters had always dreaded yet always expected.

The space shuttle Columbia had disintegrated 37 miles above Texas, seven astronauts had been killed and America's space program, always an eye blink from disaster, had just suffered its second catastrophic in-flight failure. And unlike the Challenger disaster 17 years earlier, Columbia's destruction left the nation one failure away from the potential abandonment of human space exploration. With 40 percent of the shuttle fleet now lost, a third disaster almost certainly would mean the end of the program.

Several weeks later, while flying home to Florida after one of countless trips to Houston, Cabbage and Harwood decided to join forces to tell the full story of Columbia's final mission: the human and technical causes; the historical context surrounding the accident; what it could mean for the future of U.S. space flight. This book is the result.

The original idea was to deliver a straightforward critique of the disaster and its causes as seen through the prism of the Columbia Accident Investigation Board's seven-month investigation. However, as the authors dug deeper and interviewed dozens of participants across the country, it quickly became clear there was an even more compelling story: the human drama that had unfolded on the ground in the weeks and months before the tragedy. In an effort to more fully capture that drama, the authors decided to tell the story as a chronological narrative, focusing more on what individuals knew and when they knew it, without the benefit of 20/20 hindsight.

The result is a complex tale of how smart, well-intentioned people—NASA's best and brightest—repeatedly missed the warning signs that made another shuttle disaster inevitable. There are no genuine heroes or villains here. This account, like real life, is too complex to be neatly compartmentalized into such clear-cut categories. It stands instead as a stark reminder of philosopher George Santayana's timeless adage, "Those who cannot remember the past are condemned to repeat it."

MICHAEL CABBAGE
WILLIAM HARWOOD

December 1, 2003
Cape Canaveral, Florida

ONE

RE-ENTRY

Looks like a blast furnace.
—*Shuttle commander Rick Husband, midway through re-entry*

Plunging back to Earth after a 16-day science mission, the shuttle Columbia streaked through orbital darkness at 5 miles per second, fast enough to fly from Chicago to New York in two and a half minutes and to circle the entire planet in an hour and a half. For Columbia's seven-member crew, the only hint of the shuttle's enormous velocity was the smooth clockwork passage of entire continents far below.

Commander Rick Husband knew the slow-motion view was misleading, a trick of perspective and the lack of anything nearby to measure against the craft's swift passage. He knew the 117-ton shuttle actually was moving through space eight times faster than the bullet from an assault rifle, fast enough to fly the length of 84 football fields in a single heartbeat.

And Husband knew that in the next 15 minutes, the shuttle would shed the bulk of that unimaginable speed over the southwestern United States, enduring 3,000-degree temperatures as atmospheric friction converted forward motion into a hellish blaze of thermal energy. It had taken nearly 4 million pounds of rocket fuel to boost Columbia and its crew into orbital velocity. Now the astronauts were about to slam on the brakes.

For Husband, a devout Christian who put God and family ahead of his work as an astronaut, flying this amazing machine home from space was a near religious experience in its own right, one he couldn't wait to share with family and friends gathered at the Kennedy Space Center in Florida. He had served as pilot on a previous shuttle flight, but this was his first as commander, and in the world of shuttle operations, it's the commander who actually lands the spacecraft.

He relished the opportunity. But his life as an astronaut took a backseat to his deep faith in God. Before blasting off on his second space flight as commander of Columbia, he videotaped 34 Bible lessons for his two kids, one each for the 17 days he would be away from home.

"The space shuttle is by far the most complex machine in the world," he had told his hometown church congregation three years earlier. "When you think about all the thousands of people it took to sit down and design this machine—the main engines, auxiliary power units, the hydraulics, the flight control systems, the reaction control jets, the solid rocket boosters, the external tank that fuels the main engines, the crew compartment with all the controls and all the time that was spent to put this thing together and make it work—it's to me inconceivable that you could take a look at the universe and think that it all just happened by accident.

"And inside that vehicle are seven astronauts, each one of which is more complex than this vehicle we went up in," he continued. "And God is an awesome God."

Looking over his cockpit instruments as he prepared Columbia for entry, the 45-year-old Air Force colonel chatted easily with his crew-mates, coming across more as an older brother than as the skipper of a $3 billion spacecraft. But underneath the friendly camaraderie was the steady hand of a commander at ease with leadership and life-or-death responsibility.

"People have characterized him as a laid-back guy, easy-going," said entry flight director LeRoy Cain, who shared Husband's deep religious convictions. "But a lot of that was based in his faith, realizing our time here is limited and ultimately the real goal is to have that relationship with your maker. And he had that and he wanted to share that in a way that wasn't intrusive or offensive. And that's the biggest reason this crew gelled so well together."

Husband was also the first pilot since the astronaut class of 1984 to be given a shuttle command on his second mission. Kent Rominger, chief of the National Aeronautics and Space Administration's astronaut corps and commander of Husband's first mission, said Husband "came out of that flight with a really strong reputation. Rick worked hard, did a really good job, was a great leader. He was a really gifted pilot."

So good, in fact, that data tapes charting his every move at the controls of NASA's shuttle training aircraft were frequently used to show other pilots how a textbook approach and landing should be flown.

"This is Mission Control, Houston. Columbia's altitude is now 90 miles above the Pacific Ocean to the north of the Hawaiian Islands, about two minutes away from entering the Earth's atmosphere," said NASA commentator James Hartsfield, his words carried around the world by satellite over NASA's television network. "All activities continuing to go smoothly en route toward a touchdown at the Kennedy Space Center at 8:16 a.m. Central time."

Getting to Columbia's flight deck hadn't been easy for Husband, who grew up dreaming about one day flying in space.

"I've wanted to be an astronaut all my life, ever since I was about four years old," he once said. "It was the only thing I could think about wanting to do."

So he planned his education and a military career with that single goal in mind.

After graduating from high school in 1975, the boy from Amarillo enrolled at Texas Tech University in Lubbock, a two-hour drive, where he earned a bachelor's degree in mechanical engineering in 1980. There, he fell in love with Evelyn Neely, who, like Husband, had grown up in Amarillo. The two were married in their hometown at First Presbyterian Church. Now, 20 years later, the couple had two children, a 12-year-old daughter, Laura, and a 7-year-old son, Matthew.

The first seven years after his college graduation included an endless procession of Air Force bases, where he learned to fly the F-4 fighter and eventually became so good at it he was promoted to instructor. In 1987, he was assigned to the legendary test pilot school at Edwards Air Force Base in Southern California's Mojave Desert. As an Amarillo schoolboy, he had built models of the flame-belching missiles that catapulted his heroes into orbit. Now, here at the same place where Chuck Yeager had broken the sound barrier, Husband proved he also had the right stuff.

But that wasn't enough.

By the time he arrived at Edwards, Husband already had applied to be an astronaut once and had been turned down. That was just prior to the explosion of the shuttle Challenger in 1986, and NASA ultimately canceled all new astronaut hires. Husband applied again afterward and was turned down a second time. Realizing NASA wanted astronaut candidates with advanced degrees, he went back to school at Fresno State University and earned a master's degree in mechanical engineering. The third time around, he got as far as the Johnson Space Center in Houston for a week of interviews and tests. Worried he might not pass the physical this time because of his eyesight, he wore contact lenses and lied about that on the application. He passed the physical, but again, the answer was no.

In the meantime, Husband drew an assignment as an exchange pilot with Britain's Royal Air Force in 1992 and shipped off to Boscombe Down, England, where he helped test a variety of new aircraft. He prayed for guidance on what he should do.

"God showed me that lying certainly was not the kind of thing that a Christian is supposed to do," he reflected in 2000. "When it came time for me to fill out the application a fourth time, I felt the strongest prompting from God to tell the truth. In studying the Bible more, I had come across Proverbs 3:5–6 that says, 'Trust in the Lord with all your heart and lean not on your own understanding; in all your ways acknowledge him and he will make your paths straight.'

"It was as if God was saying 'Just trust me! You lied last time and didn't make it. Try telling the truth this time and see what happens.' Finally, I had come to the point where I understood what it meant to give my life to God and to trust Him. I said, 'OK, Lord. I want to do what You want me to do, and it doesn't matter if I'm an astronaut or something else.' "

This time he told the truth and was invited to begin training at Johnson in December 1994. But despite his obvious skills in the cockpit, it would be another five years before he was assigned to his first mission. Only two other members of the 20-member astronaut class known as Group 15 waited longer.

"A lot of people didn't know that Rick was one of the last people to fly in his class," said Dave Pitre, an astronaut trainer assigned to Columbia's mission. "And I don't know the reason why, because I think Rick's a great guy. But this is what I speculate.

"You look at his résumé and he's the best. But he's so unassuming. And on the sixth floor [of the astronaut office], there's a lot of competition to be the best, to do the best. There's a lot of jockeying. Rick wouldn't do that, he never did that. He let his actions speak for himself. So he just sat there patiently in line and waited, and waited, and waited. I think he just wanted to make a statement and so they all tried just so hard, I mean, every training session you just got the feeling these guys were putting out 110 percent all the time. Everybody had something to prove on that flight and I think it just bonded them all together."

"Columbia is currently targeted toward runway 33 at the Kennedy Space Center; the runway selection continues to be dis-

cussed here in Mission Control, however. But for its approach to runway 33, Columbia will perform a right overhead turn to align with the runway of about 214 degrees."

Preparations for the shuttle's arrival at Kennedy were in full swing. The landing support team, made up of engineers and technicians required to deactivate critical systems after touchdown, had gathered at dawn to go over their plans and to prepare their equipment. They were stationed at the northwest end of the broad, 3-mile-long Shuttle Landing Facility runway, expecting Columbia to come in from the southeast.

Shortly after sunrise, buses, limousines, and a fleet of sport-utility vehicles and government vans brought a crowd of VIPs, NASA managers, reporters, and invited guests to bleachers strung out on the eastern side of the Shuttle Landing Facility, about midway down the runway. Hartsfield's voice blared from loudspeakers mounted on telephone poles behind the bleachers, sounding clear in some areas but muffled in others. TV monitors, carrying NASA's television coverage of the landing, were spaced out in front of the bleachers, and a large countdown clock was ticking down toward touchdown.

The astronaut families were gathered at a lone set of bleachers at the northernmost end of the midfield viewing site, cordoned off from the other VIPs and guests to ensure a bit of privacy. An astronaut was assigned to each family to answer questions and to provide assistance in case of an emergency.

No problems were expected, and a triumphant homecoming was just minutes away. It was 8:44 a.m. on Feb. 1, 2003, and Columbia was descending through 400,000 feet northwest of Hawaii.

"OK, we're just past EI," Husband told his crewmates, marking when Columbia, flying wings level, its nose tilted up 40 degrees, finally fell into the discernible atmosphere.

He was referring to "entry interface," the moment the shuttle descended through an altitude of 76 miles. At that altitude—11 times higher than a typical passenger jet flies—the atmosphere is still a vacuum in the everyday sense of the word. But enough atoms and molecules are present to begin having a noticeable effect on a vehicle plowing through them at 25 times the speed of sound.

Wearing bulky, bright orange pressure suits, Husband, rookie pilot William "Willie" McCool, flight engineer Kalpana Chawla (pronounced KULP´-nah CHAV´-lah), and Navy physician-astronaut Laurel Clark were strapped into their seats on Columbia's cramped flight deck, working through the final entries on a long checklist.

The shuttle's flight computers, each one taking in navigation data and plugging the numbers into long strings of equations, were doing the actual flying. Husband wouldn't take over manual control until the orbiter was on final approach, 50,000 feet above Kennedy. During this phase of entry, the astronauts were monitoring the ship's progress, discussing the view outside and making last-minute adjustments to their pressure suits. Husband and McCool had just finished drinking a final few bags of salty water in a somewhat unpleasant procedure known as "fluid loading." The concoction would make them less susceptible to feeling woozy during the onset of gravity after 16 days in weightlessness.

Husband was in the front left seat, the command position aboard any aircraft, with McCool to his right on the other side of a switch-studded instrument console. Chawla, a native of India, was a veteran of one previous shuttle flight and an accomplished pilot. Something of a legend in her hometown of Karnal in the Indian State of Punjab, Chawla was a role model in a country where less than half the women were literate. She sat directly behind the central console, calling out and double-checking re-entry tasks. Clark was seated to Chawla's right, almost touching shoulders with the diminutive flight engineer.

Strapped into seats on the split-level crew cabin's lower deck were payload commander Michael Anderson, another shuttle veteran and one of only a handful of African-American astronauts at NASA, physician-astronaut David Brown, a former circus acrobat, jet pilot and amateur videographer, and fighter pilot Ilan Ramon, the first Israeli to fly in space.

The crew was well aware of the high-risk nature of a shuttle flight. Like all astronauts, they had put their affairs in order before flying to Florida for the launch. Anderson, who attended the same Houston church as Husband, typified the tightly knit crew's feelings about personal safety. He told a former pastor not to be concerned if he didn't make it back from a mission. "Don't worry about me. I'm just going on higher."

High-risk missions were nothing new for Ramon. An impressive figure in the Israeli air force, he helped lead a daring 1981 Israeli bombing raid that reduced an unfinished Iraqi nuclear reactor to rubble. But he did not join Columbia's crew as a warrior.

"I represent, first of all, of course the state of Israel and the Jews," he said during an orbital interview a few days earlier. "But I represent also all our neighbors. And I hope it will contribute to the whole world and especially to our Middle East neighbors."

Unlike the upper flight deck with its wraparound airliner-type cockpit windows and large overhead view ports, the lower deck featured a single, small porthole in the shuttle's main hatch, almost out of view on the left side of the cabin. For Anderson, Brown and Ramon, there was nothing to see but rows of equipment lockers. At least they were plugged into the ship's intercom system, following along as the flight deck crew worked through the re-entry checklist.

They were listening in a half-hour earlier as Husband counted down to deorbit ignition, when Columbia's flight computers fired up the shuttle's twin braking rockets as the spacecraft flew upside down and backwards 170 miles above the central Indian Ocean. The two-minute 38-second rocket firing slowed the shuttle by just 176 mph. But that small decrease was just enough to lower the far side of Columbia's orbit deep into the atmosphere above Florida's east coast.

For the first half hour of re-entry, Columbia and its crew simply fell through the black void of space on a precisely plotted course toward a runway on the other side of the planet. But now, finally back in the discernible atmosphere, things were about to get interesting.

For McCool, an accomplished Navy carrier pilot and father of three who brought a boyish enthusiasm to Columbia's flight deck, entry interface was a long-anticipated milestone. Veteran astronauts had told him to expect a spectacular light show. Right on cue, the inky blackness outside his cockpit windows began giving way to a faint salmon glow.

At first, the effect was so subtle he wasn't sure it was really there.

"That might be some plasma now," he observed, as if seeking confirmation from Husband.

"Think so already?" asked Clark, seated directly behind McCool and aiming a handheld video camera out the cockpit's overhead windows.

"That's some plasma," Husband finally confirmed.

"Copy, and there's some good stuff outside," Clark replied. "I'm filming overhead right now."

"It's kind of dull," McCool said a moment later, as if disappointed.

"Oh, it'll be obvious when the time comes," Husband reassured him.

McCool didn't have to wait long. During the next minute, the initial faint glow steadily brightened until there was no mistaking that the shuttle was embedded in a fireball.

"It's going pretty good now, Ilan," McCool reported to the three fliers on the lower deck. "It's really neat, just a bright orange-yellow out over the nose, all around the nose."

"Wait until you start seeing the swirl patterns out your left and right windows," Husband said.

"Wow."

"Looks like a blast furnace," Husband said dryly.

Even so, he and his crewmates were not worried. Focused, yes. Aware of the danger, yes—as any well-trained astronauts would be when contemplating the energies involved in a shuttle launch or re-entry. But not worried. After all, there was no reason for concern. In the 111 previous space shuttle re-entries, there never had been a catastrophic "inflight anomaly," as NASA refers to out-of-the-ordinary events. The only disaster in the history of the program, the Challenger explosion, had occurred during launch when one of its boosters had ruptured.

The Columbia astronauts already had survived the eight-and-a-half-minute climb to orbit that most experts, with some reason, considered a far more dangerous phase of flight. A fully fueled space shuttle weighs 4.5 million pounds at liftoff, yet accelerates to 100 mph—straight up—in about 10 seconds. In the first minute, more than two million pounds of fuel are consumed by the ship's twin solid-fuel boosters and three hydrogen-fueled main engines as the spacecraft plows its way through the dense lower atmosphere. Seven and a half minutes later, the astronauts are in orbit, moving through space at more than 17,000 mph.

Compared with launch, getting home was a walk in the park. Or so it seemed to most, including many at NASA. The astronauts knew better, of course. All of the energy it took to boost Columbia to orbital speed was still there in the form of the craft's enormous velocity. To make it back to Earth, the shuttle would have to give up that energy in the form of heat from atmospheric friction.

But no one was particularly worried. Of NASA's four space shuttle orbiters—Columbia, Discovery, Atlantis and Endeavour—Columbia had a well-earned reputation for experiencing most of its problems on the ground. If the launch team could just get the countdown to zero, engineers joked, NASA's oldest space shuttle would chalk up a smooth flight and a trouble-free return to Earth.

There were exceptions, of course. In 1999, a short circuit five seconds after liftoff knocked out one of Columbia's three main electrical circuits, shutting down two computers that controlled the main engines and leaving the ship one glitch away from a potentially catastrophic failure. Frayed insulation on old wiring turned out to be the culprit, raising questions about the general health and well-being of NASA's aging fleet of space shuttles.

But the wiring problems were corrected, and Columbia, making its second flight since a major overhaul, was completing another near-perfect mission, returning to Earth after a 6.6-million-mile voyage spanning 255 orbits since blastoff.

In Mission Control, Cain and flight dynamics officer Richard Jones were discussing the weather in central Florida and trying to decide which end of the shuttle's runway Husband should use. From 50,000 feet on down, Husband would be flying Columbia manually, and Cain wanted to guarantee the commander the best possible conditions.

Like Husband, the 38-year-old Cain lived a life that revolved around family, church, and space exploration. He earned a bachelor's degree from Iowa State University in 1988 and promptly went to work for Rockwell Shuttle Operations in Houston as a guidance systems officer in Mission Control.

NASA hired him in 1991, the year the first of his three daughters was born, and he quickly proved his mettle. In 1998, he was selected as a flight director, becoming one of the chosen few entrusted with overseeing operations during the shuttle's high-risk climb to orbit and return to Earth. In the mission operations world of ascent and entry, the shuttle commander and the flight director work hand in hand, one in charge on orbit, the other in charge on the ground. The success of a mission rests squarely on their shoulders. And at no time are the stakes higher than during launch and landing.

"Between you and me, I am nervous every time I walk in the door," Cain said. "People have asked me, is it scary? Hell yeah! Every time I walk in the door, it's almost like—and this sounds corny, but this is what happens to me—something comes over me. It's just like this feeling, and suddenly it's almost like this instantaneous rush of focus."

For Columbia's entry, Cain began his day with a standard routine. After a handover briefing from the off-going flight control team, Cain gave the entry team a brief weather update, assessing the odds that Mother Nature would cooperate with their plans to bring Columbia home that morning. He closed with a familiar message: "Let's go get 'em, guys."

The communications loops in Mission Control were quiet as Columbia continued its automated descent under the control of its four primary flight computers, approaching the coast of California just north of San Francisco. Mechanical systems officer Jeff Kling, monitoring telemetry from the shuttle's myriad systems, saw nothing unusual in the numbers flickering on his computer displays. Columbia was living up to its reputation.

To Cain's right, looking on through a large window on the second floor of the Mission Control center were Johnson's senior shuttle managers, including program manager Ron Dittemore, Mission Management Team chairwoman Linda Ham and shuttle engineering director Ralph Roe. NASA's senior astronaut, John Young, was there along with Frank Benz, director of engineering at Johnson. They were all listening to the chatter on the flight control loops, monitoring Columbia's descent.

Seated directly behind Cain were veteran flight directors John Shannon and Phil Engelauf, along with astronaut Ellen Ochoa. All three were on hand to provide advice or assistance if necessary. Hartsfield was seated to Cain's left in the back corner of the room, listening to the loops and relaying details of Columbia's descent to the public.

But there were no issues of any significance to discuss. On a huge projection television screen at the front of the room, Columbia's actual location was plotted against the ship's predicted path on a map showing the shuttle's planned route to Florida. With a glance, anyone in the room could see the shuttle was right on course.

The only problem of any significance reported by NASA during Columbia's 28th mission—and it was more an annoyance than anything else—was the failure of a humidity control system in the crew's Spacehab experiment module. The glitch forced the astronauts to shut down one of their air conditioners early on, and temperatures in the research module went up slightly as a result. But the crew had no complaints. After a wait of two and a half years for mission STS-107 to get off the ground, a bit of unexpected heat was inconsequential.

The growing heat outside Columbia, of course, was another matter.

"This is amazing. It's really getting, uh, fairly bright out there," McCool observed as the glow around the orbiter continued to intensify.

"Yeah, you definitely don't want to be outside now," quipped Husband.

"What, like we did before?" Clark said pointedly as her crewmates laughed.

"Good point," Husband replied.

It was 8:47 a.m. and just three minutes had passed since entry interface. Approaching the coast of northern California, Columbia was dropping like a rock, its nose-up orientation designed to focus reentry temperatures of up to 3,000 degrees on the heat-resistant reinforced carbon-carbon [RCC] panels making up the wings' leading edges and the orbiter's nose cap. Thousands of black heat-shield tiles making up the belly of the spacecraft would protect the underlying, vulnerable, aluminum skin from slightly lower, but still extreme, temperatures.

The spacecraft had not yet slowed much—its velocity was still a blistering 24.7 times the speed of sound—but aerodynamic pressure was steadily building across the underside of the shuttle. It was now up to a half pound per square foot as the craft continued its descent through the thin upper atmosphere. In another minute, it would quadruple, and one minute after that, it would be 20 times greater.

Temperatures on the nose and the wing leading edges already were above the 1,200-degree melting point of aluminum.

> *"Columbia's course toward Florida will take it across the continental United States, crossing the California coast above the San Francisco Bay area and continuing across Sacramento, California, providing a spectacular view for persons in that area of Columbia's descent through the atmosphere."*

A continent away at Kennedy, excitement was building for Columbia's homecoming.

When the first members of the landing support team had begun arriving at 6 a.m., the Shuttle Landing Facility, along with the rest of Cape Canaveral, was buried beneath a thick fog. Now, as Columbia zipped toward the California coast, the fog had burned off and given way to a glorious midwinter morning at the Cape, with 52-degree temperatures, light westerly winds, and a few scattered clouds—near-perfect conditions.

Only an hour earlier, questions about how quickly the fog would lift had presented Cain and the flight control team with a minor dilemma: Do we bring Columbia home on its 255th orbit around Earth? Or do we wait another hour and a half for a try on the 256th revolution, Columbia's final landing opportunity of the day? Waiting an orbit ran the risk of clouds forming near the Cape that could obscure the shuttle's runway.

Ham's Mission Management Team had decided the day before not to activate the shuttle's backup landing site at Edwards. Landings in California added a week or more to a shuttle's ground processing schedule, and it cost NASA about $1 million to fly a shuttle back to Florida atop a 747 transport jet.

As a result, Cain had two shots at getting Columbia back to Florida on Feb. 1 or the mission would have to be extended 24 hours. That's how long it would take before the shuttle's orbital position and the Kennedy landing site synched back up again.

Carved out of a mosquito-infested bog on the edge of the Indian River, the shuttle's 15,000-foot runway was about a third longer and twice as wide as those at major international airports. Columbia and its sister shuttles, however, returned to Earth as huge, unpowered gliders. The joke was they flew like bricks and dropped like rocks. Unlike airplane pilots, shuttle commanders couldn't pull back on the stick and open up the throttle to fly around for another approach if something went wrong. From the moment the braking rockets completed the deorbit burn, Columbia and its crew were locked into a landing at Kennedy. That made the weather a big deal.

Overhead, Rominger was flying a NASA business jet modified to handle like a space shuttle on final approach to the runway. He was making practice approaches to both ends of the runway to check visibility and turbulence.

Joining Rominger aboard the Shuttle Training Aircraft that day was Barbara Morgan, an Idaho school teacher who had served as Christa McAuliffe's backup in NASA's original Teacher in Space Program. Morgan was at Kennedy 17 years earlier, on Jan. 28, 1986, when McAuliffe and her six crewmates died in the explosion of the space shuttle Challenger.

Morgan never gave up her dream of fulfilling McAuliffe's legacy. Now, she was back as a full-fledged astronaut, a member of the crew assigned to Columbia's next flight. She was "shadowing" Rominger, accompanying him through a complete launch-and-landing cycle as part of the normal on-the-job training for a new astronaut. Morgan had come full circle in her quest to fly in space.

Rominger reported both ends of the runway were acceptable for use, and Cain earlier had tentatively settled on runway 33, which meant Columbia would be touching down on the landing strip's southeast end and rolling to the northwest. But he was still considering a switch to the other end to keep the sun out of Husband's eyes on final approach. The photographers and local television crews hoped he stuck with his original decision. Runway 33 meant a better view of the landing.

Workers in the landing support team, gathered at the northwest end of the Shuttle Landing Facility, also were hoping Mission Control would stick with 33. Otherwise, they would have to hurriedly reposition the slow-moving caravan of 20 or so vehicles, known as the Orbiter Recovery Convoy, to the other end of the runway. The technicians' jobs ranged from helping the crew off the shuttle to hooking up cooling systems and making hazardous systems safe.

Waiting with eager anticipation was Ann Micklos, a senior engineer and an expert on the shuttle's heat-shield tiles. She had been dating Brown. While the relationship had cooled, they were still close friends, and she could hardly wait to welcome him home. He had first asked her out in July 1999, when both were on this same runway after another landing by Columbia. Today, he was returning to Earth aboard the same orbiter, carrying a Swiss watch he had bought her as a birthday present the previous June. She never had seen it. It was going to be a surprise.

> *"It'll be visible as well through much of the United States' southwest above southern Nevada and northern Arizona and central New Mexico as it continues its descent through the atmosphere, trailing a plasma trail left as it heats the atmosphere around it during its descent."*

Aboard Columbia, the astronauts were putting on their gloves, pressurizing their spacesuits and conducting routine communications checks to make sure they could hear each other with their helmet visors down and locked. Clark, who planned to videotape the entire descent, was fiddling with her camera, taking occasional shots through the overhead windows as the plasma continued to build around the vehicle.

"Willie, I can see you in your mirror," Clark said playfully, looking over the pilot's shoulder at a small mirror attached with Velcro to the forward dashboard.

"Now I can see your camera!" McCool said to Clark, looking back in the mirror.

"OK," Husband said, his tone of voice gently telling his crewmates to cut out the horseplay.

"Stop playing," Chawla acknowledged with good humor.

Just before 8:50 a.m., still off the coast of California, Columbia's computers ordered small, right-side steering rockets to fire, moving the shuttle's nose to the right. These roll maneuvers were planned to bleed off velocity before reaching the landing site.

McCool was looking forward to getting a little flying time in the next few minutes. After receiving the go-ahead to fire Columbia's braking rockets to begin the trip home, Husband had notified Mission Control that McCool would be taking the stick briefly, just before touchdown, as the shuttle banked to line up on the runway. Not every commander gave his or her co-pilot a chance to actually fly the shuttle, but that was Husband's nature. McCool couldn't wait.

> *"Columbia in almost an 80-degree-bank to the right to dissipate speed, the first of four banks it performs as it approaches Florida to slow down as it descends. Altitude now 47 miles or about 248,000 feet. The shuttle's speed is 16,400 miles per hour."*

At 8:53 a.m., 23 minutes from touchdown, Columbia finally crossed the coast of California, a long plume of super-heated air trailing in its wake. On the big map in Mission Control, the red triangle marking the shuttle's position crept on to land north of San Francisco. Scores of shuttle watchers and amateur photographers across the Southwest had gotten up early to witness Columbia's fiery descent. Bill Hartenstein, a veteran shuttle photographer, and Gene Blevins, a freelance photographer who worked with the *Los Angeles Daily News,* had set up six cameras at the Owens Valley Radio Observatory near Bishop, California, figuring a shot of Columbia streaking through the pre-dawn sky above the big radio telescope dishes in the foreground would be spectacular. Blevins was not disappointed.

"Once we got our cameras set up, sure enough, around 5:52, 5:53 [a.m. Pacific time], here comes this big white dot out over the mountains coming right at us," Blevins said. "This thing was coming in at incredible speed."

He and Hartenstein knew the pictures would, in fact, be spectacular. The conditions were perfect, and in a time-exposure photograph, Columbia would show up as a brilliant streak across the dark sky. But as he watched the shuttle race by, Blevins realized the plasma trail didn't look quite right. It was a subtle effect, and he didn't immediately think anything of it as he tended his cameras. But as the shuttle moved away and the camera shutters closed, "I saw this big red flare come from underneath the shuttle and being forced downward. I was like, 'whoa!' " He turned to Hartenstein. "Bill, did you see that? Something came off the shuttle!"

Hartenstein looked up from his cameras. He didn't see anything unusual, just the dimming space shuttle disappearing from view toward the horizon.

"Shuttle's altitude now 45 miles, speed 15,800 miles per hour, continuing in a right bank with wings angled 70 degrees, the first of four banks it performs to dissipate speed as it approaches landing."

Activity was picking up at Kennedy. At the midfield viewing site, Ramon's wife, Rona, and the couple's four children waited with nervous anticipation. Matthew Husband ran around playing under the watchful eyes of mother Evelyn and older sister Laura. McCool's wife, Lani, and two of their three sons—Christopher and Cameron—were there, along with Anderson's wife Sandra, their two young daughters and Chawla's husband, Jean-Pierre Harrison. Brown's parents, at home in Virginia, were watching the entry on NASA television as was McCool's oldest son, Sean, at college back in Canada.

"We're just standing around, it's a beautiful, kind of pleasant day," said Clark's husband, Jon, a Navy captain and NASA flight surgeon. "There wasn't a cloud deck forming, winds weren't at crosswind limits, you know, all that kind of stuff. So I'm thinking hey, we're good to go."

The broad runway was familiar territory for Clark. Years earlier, he had arranged for her to play the role of an injured shuttle flier in a landing disaster simulation at Kennedy. Now she was returning from her first flight, to the very spot where she had once pretended to be a casualty.

The couple's 8-year-old son, Iain, eagerly awaited the sonic booms that would herald Columbia's approach. He, his parents and the family dog had survived a harrowing crash the previous December, when Jon's single-engine Bonanza airplane hit a violent downdraft while trying to land during a storm. No one was injured, but the plane was destroyed. The experience haunted young Iain, and he begged his mother not to fly on the shuttle.

"He was very worried, very worried about her," Jon recalled. "He had some very, very moving premonitions that something bad was going to happen and he didn't want her to go."

In a family video conference during Columbia's flight, Iain asked his mother "Why did you go?" But today, all was forgiven as he romped on the grass in front of the bleachers.

And it wasn't just the children who were happy. A joyful mood had settled in across the midfield viewing site. Just on the other side of the yellow rope separating the astronaut family bleachers were top NASA officials and Kennedy staff.

"I can hear the loudspeakers going but I'm not paying any attention to them because I'm just talking to people and it's kind of a festive atmosphere there," said Wayne Hale, a veteran flight director on assignment to Kennedy for a year to oversee shuttle processing. "I have no real responsibility [for landing]. We're out there, we're talking about, you know, how we're going to walk around the vehicle and greet the crew and how soon can we do that after they land, that sort of stuff."

At the south end of the viewing site, separated by a small construction zone where NASA was building a new control tower, reporters from a dozen or so news organizations milled about, taking notes, checking the weather and shooting the breeze. Media interest in landings had waned in recent years. But this morning, the reporters' ranks were slightly swelled by a small influx of Israeli journalists there to document Ramon's historic return. As usual, a handful of satellite trucks from Orlando's network television affiliates were standing by, ready to briefly interrupt cartoons and other Saturday morning fare to broadcast Columbia's arrival.

Several reporters were killing time by joking with Bill Johnson, the white-haired, pony-tailed, veteran news chief of Kennedy's public affairs office. One journalist lamented having to get up early on a weekend morning to see an event that had become so routine it no longer was newsworthy.

The cautious Johnson replied that the winds had slightly shifted since Columbia fired its maneuvering engines to come home. In addition, he noted, the onboard laboratory in the orbiter's cargo bay made Columbia's landing weight ever-so-slightly heavier than normal. A reporter next to Johnson piped up with a reminder that Husband was a rookie commander who would be landing the shuttle for the first time and his co-pilot, McCool, never had flown before.

"Oh well," another reporter joked, "I guess that all means they're doomed for sure."

Everyone laughed.

At about that same moment, as Columbia streaked toward the California-Nevada border more than 2,000 miles away, mechanical systems officer Kling in Mission Control noticed something unusual in the stream of telemetry from the space shuttle. Downward-pointing arrows appeared beside readings from two sensors measuring hydraulic fluid temperatures in the shuttle's left wing.

"What in the world?" another mechanical systems engineer, seeing the same data in a different room, said to Kling.

"This is not funny," Kling replied. "On the left side."

"On the left side," the engineer agreed.

They both tried to find some common thread that might explain the readings. Then, a few seconds later, two more down arrows appeared. It was as if the wiring to the four sensors in question, in hydraulic lines leading to the wing's flaps, or elevons, had been cut. Kling notified Cain.

"FYI, I've just lost four separate temperature transducers on the left side of the vehicle, hydraulic return temperatures," Kling reported, speaking quickly. "Two of them on system 1 and one in each of systems 2 and 3."

"Four hyd return temps?" Cain calmly asked.

"To the left outboard and left inboard elevon."

"OK, is there anything common to them? . . . I mean, you're telling me you lost them all at exactly the same time?"

"No, not exactly," Kling said. "They were within probably four or five seconds of each other."

"OK," Cain said, mulling over possible explanations. "Where are those, where is that instrumentation located?"

"All four of them are located in the aft part of the left wing, right in front of the elevons, elevon actuators," Kling replied. "And there is no commonality."

Cain pondered that for a moment.

"No commonality," he said after a long pause, his tone of voice indicating bafflement. Multiple sensor failures were rare, usually the result of problems with some common electrical system or component shared by the sensors in question. But these data "dropouts" could not immediately be traced to some single failure point.

Cain's thoughts flashed back to Columbia's launch. He recalled a briefcase-sized piece of foam insulation that broke away from Columbia's external fuel tank 81 seconds after liftoff and slammed into the underside of the left wing. A team of experts had studied the impact and dismissed it as not being a safety-of-flight issue. The Mission Management Team had unanimously accepted the results of the analysis.

> *"Columbia continuing in a right bank, the wings angled 43 degrees, speed 15,000 miles per hour, altitude 43 miles, 2,090 miles to touchdown at the Kennedy Space Center targeted for runway three-three at Kennedy at present. Crossing the continental United States, now crossing above southern Nevada to the north of Las Vegas."*

An unsettling thought crossed Cain's mind. Was the loss of four left-wing temperature sensors related to the foam strike? It couldn't be, he thought. That had to be a coincidence.

"Tell me again which systems they're for?"

"That's all three hydraulic systems," Kling said. "It's . . . two of them are to the left outboard elevon and two of them to the left inboard."

"OK, I got you," Cain said. It was 8:56 a.m. and Columbia was crossing the Utah–Arizona–New Mexico state lines. Aerodynamic pressure was up to 40 pounds per square foot as the shuttle continued its steep descent.

In the management viewing room above Mission Control, Ham was also worried. About the left wing. About the foam strike earlier in the mission. In her pivotal role as chairwoman of the Mission Management Team, Ham had approved the analysis that concluded the foam strike was not a safety of flight issue. She turned to Roe, a tall ex-football player, who also had taken part in the deliberations.

"Ralph, it's the left wing."
"It's not that," he said.

"Columbia's course continuing across Arizona and the Arizona and New Mexico border near the Four Corners area of the United States. Its course will take it almost directly above Albuquerque, New Mexico, its altitude now 225,000 feet or 42 miles, speed 14,300 miles per hour, 1,785 miles to touchdown at the Kennedy Space Center. It's banking now back to the left, the second in a series of four banks that dissipate speed of the spacecraft as it becomes an aircraft and descends into the atmosphere toward Florida. Wings angled about 75 degrees to the left."

Other than the four temperature sensors on the hydraulic fluid return lines, everything appeared normal in Mission Control. Flight directors, however, are trained to worry, and Cain was no exception. He told reporters the day before he wasn't concerned about the foam strike, that it had been analyzed and that engineers did not think it was a safety-of-flight issue. But Cain didn't like coincidences.

"My very first thought, when he said it was on the left wing, was we had hot gas in the wing," Cain said later. "I almost pictured the whole sequence in my mind in the first few microseconds. It was pretty chilling. But I didn't believe it."

Cain called guidance, navigation, and control officer Mike Sarafin for a second opinion.

"Everything look good to you? Control and rates and everything is nominal, right?"

"Control's been stable through the rolls that we've done so far," Sarafin replied. "We have good trims. I don't see anything out of the ordinary."

Cain turned back to Kling, saying "All other indications for your hydraulic system indications are good."

"They're all good," Kling confirmed. "We've had good quantities all the way across."

"And the other temps are normal?"

"The other temps are normal, yes sir."

"And when you say you lost these, are you saying that they went to zero or off-scale low?"

"All four of them are off-scale low," Kling said. "And they were all staggered. They were, like I said, within several seconds of each other."

Off-scale low meant the sensors had simply stopped working. The shuttle was equipped with thousands of sensors, and random failures were not unusual. Nothing else in the continuous stream of data from Columbia indicated any signs of trouble. In just three minutes, the shuttle would be out of the region of maximum aerodynamic heating. There was reason to hope that all was well.

Suddenly, Husband called down, his first query since Columbia had entered Earth's atmosphere 15 minutes earlier.

"And, uh, Hou[ston] . . ." he began. His transmission was cut off. Such dropouts were not unusual during re-entry as the shuttle banked left and right, its big vertical tail fin occasionally blocking signals from reaching a NASA communications satellite stationed over the western Pacific.

A few seconds later, Kling saw more down arrows appear on his computer screen, this time signaling a loss of data from the shuttle's left main landing gear tires.

"We just lost tire pressure on the left outboard and left inboard, both tires," he told Cain, a half minute after Husband's interrupted call. The mechanical systems officer had a sinking feeling in the pit of his stomach.

Astronaut Charles Hobaugh, the flight controller responsible for talking directly to the shuttle crew, heard Kling's report to Cain and promptly radioed Husband: "And Columbia, Houston, we see your tire pressure messages and we did not copy your last."

At the runway viewing site, Jon Clark heard Hobaugh's call over the loudspeakers. A chill settled over him.

"The tire pressure call is a major off-nominal call. So I'm thinking, OK, if that's gone, they can't land with a blown tire, especially with that landing weight. And so I'm thinking, that's not good. I wonder if they're going to have to do a bailout now? And Laurel had 25 parachute jumps, she was probably more experienced than anybody as far as that goes. . . ."

No one else at the viewing site appeared to pick up Hobaugh's call to the crew. Or if they did, they didn't immediately make anything of it. Such things usually turned out to be instrumentation glitches.

But back in Mission Control, Cain was getting seriously worried. There was growing concern Columbia had suffered some sort of major malfunction that could affect its landing, now just 17 minutes away.

"Is it instrumentation?" he asked Kling, hoping the tire pressure reading was the result of a faulty sensor. "Gotta be . . ."

"Those are also off-scale low," Kling replied.

Seconds later, Husband made another attempt to contact Mission Control, replying to Hobaugh with "Roger, uh, buh . . ." He might have been saying "before" or "both," but again, the transmission was suddenly cut off and, along with it, the flow of data from the shuttle.

It was 8:59:32 a.m. Cain waited for radio contact to be restored, trying to sort out what might be wrong. Whatever it was, flight controllers would have only a few minutes or so to resolve it before Columbia reached the Kennedy Space Center.

"And there's no commonality between all these tire pressure instrumentations and the hydraulic return instrumentations?" he asked Kling.

"No sir, there's not. We've also lost the nose gear down talkback and the right main gear down talkback," the mechanical systems officer replied, referring to landing gear position sensors.

"Nose gear and right main gear down talkbacks?" Cain repeated.

"Yes, sir."

Cain's discussion was not carried on NASA's television broadcast, and as far as the public knew, everything was proceeding smoothly. Communications dropouts typically lasted several seconds, occasionally longer, but sooner or later communications always resumed. Even so, at the Kennedy midfield viewing site, Bill Readdy, NASA's associate administrator for space flight and the man ultimately responsible for shuttle operations, started paying attention.

Much of the crowd in Florida was oblivious to the drama unfolding in Mission Control. But a couple of reporters huddled around a television set inside the runway's public affairs building had noticed something peculiar.

The big map displayed in Mission Control and broadcast on NASA Television showed Columbia's progress as the ship sped across the continent toward Florida, the red triangle marking the shuttle's position. Inexplicably, that triangle had stopped moving over Central Texas.

"Columbia out of communications at present with Mission Control as it continues its course toward Florida. Fourteen minutes to touchdown for Columbia at the Kennedy Space Center. Flight controllers are continuing to stand by to regain communications with the spacecraft."

Cain queried Laura Hoppe, instrumentation and communications officer, asking how long she expected intermittent contact.

"We were rolled left last data we had and you were expecting a little bit of ratty comm, but not this long?"

"That's correct," she replied. "I expected it to be a little intermittent. And this is pretty solid right here."

Hobaugh radioed Husband to check whether the crew could hear communications from Houston.

"Columbia, Houston, comm check," Hobaugh called at 9:03 a.m. "Columbia, Houston, UHF comm check."

There was no reply.

"CAPCOM Charlie Hobaugh calling Columbia on a UHF frequency as it approaches the Merritt Island tracking station range in Florida," Hartsfield explained, his voice betraying uncertainty for the first time. "Twelve and a half minutes to touchdown according to clocks in Mission Control."

"Columbia, Houston, UHF comm check," Hobaugh tried again. It was 9:04 a.m.

The shuttle now had been out of communication for nearly five minutes. At Kennedy's shuttle landing strip, a small flurry of activity had begun in the VIP area. Clusters of people had grouped together. The laughter and smiles of just 10 minutes earlier had all but disappeared, erased by anxious looks of concern.

At the midfield park site, a landing support staging area about 100 yards in front of the VIP bleachers, NASA engineer Bob Page was standing by with a camera operator, part of a team spread out around the runway to document Columbia's landing. Page had been one of the first to review film and video of the foam impact during launch 16 days earlier.

Listening to an internal audio loop used by members of the landing photo team, Page heard that communications with Columbia had been lost. He immediately thought about the foam strike. He dropped his head, turned around, and walked off the camera mound.

At the north end of the runway, astronaut Jerry Ross, a senior manager with seven shuttle flights to his credit, was chatting with Bob Cabana, a veteran shuttle commander currently serving as director of Flight Crew Operations at Johnson. They were standing beside Kennedy's new convoy commander's van, a mobile command post loaded with communications gear and support equipment.

Suddenly, someone inside started rapping on the window and motioning for the astronauts to come back inside.

"So Bob and I went back in and the first thing they said was, 'We've lost communications with the orbiter,' " Ross remembered. "And we thought, well you know, that's not that big of a deal. And then they said, 'We've lost data.' And then we said, 'Well, if you're going to lose comm, you're going to lose data. So no big deal.' And then shortly, that was followed by 'We've lost tracking.' "

Controllers were expecting tracking radars near Kennedy to lock onto Columbia at any moment as the shuttle approached Florida. By now, they knew something was very wrong. Columbia had been out of contact longer than anyone had expected.

> *"Flight controllers are standing by for Columbia to move within communications range of the Merritt Island tracking station in Florida to regain communications."*

"When were you expecting [radar] tracking?" Cain asked the flight dynamics officer at 9:05 a.m.

"One minute ago."

At the shuttle runway, cell phones starting ringing. Soon, it appeared half the people in the crowd had phones pressed against their ears. Sean O'Keefe, NASA's administrator, heard Readdy say "This is not right, something is not right on this." O'Keefe was stunned.

Readdy, the veteran shuttle commander and former fighter pilot, was visibly trembling, his face ashen.

"It was just one of those hit-you-with-a-mackeral kind of moments," O'Keefe said. "You know, 'Good God almighty.' You see someone like him sitting there doing that, and he knows the gravity of it better than anybody. It was enough to make you just start shaking right down to the edge."

Slowly, word began to ripple through the gathering of reporters to the south that something might be amiss. Groups began to form around loudspeakers and television sets. Some reporters began to wonder aloud in hushed tones what could account for the lengthy communications dropout.

Others, unable to hear Hartsfield's commentary, kept watch on the big countdown clock, waiting with mounting concern for the twin sonic booms that would herald Columbia's arrival.

Readdy was carrying a detailed disaster plan in a thin notebook, a set of instructions worked out in the wake of the Challenger disaster outlining what needed to be done in the event of a launch or entry mishap. He began telling other senior managers to get moving, to rendezvous back at the Launch Control Center.

Kennedy shuttle manager Mike Wetmore was overwhelmed. He turned to look at the astronauts' families, the kids still running around in front of the bleachers, unaware of the sense of dread settling over the managers. He was paralyzed by the sight.

"The families were celebrating and happy and I just . . . I couldn't move. They had no idea," Wetmore said. "I am a dad, I am a dad with two small kids and I am looking over there at these little kids running around, playing and laughing. I just remember a couple of small kids chasing each other back and forth. I was just standing there frozen."

"Columbia, Houston, UHF comm check," Hobaugh radioed again.

There was no reply.

T W O

PREPARATIONS

They bonded so well. They actually loved each other like family.
—*Darla Racz, astronaut training manager*

One day in late 1995, Jeremy Issacharoff, a former political counselor at the Israeli Embassy in Washington, took his 5-year-old son, Dean, to the National Air and Space Museum. As the two strolled through the crowded exhibits, ambling past Neil Armstrong's Apollo 11 capsule and under Chuck Yeager's bright-orange rocketplane, Dean stopped at a space shuttle display. He noticed that dozens of non-U.S.

astronauts, including a member of the Saudi royal family, had flown aboard a shuttle.

Dean looked up at his father and asked the obvious question: "Daddy, why isn't there an Israeli astronaut?"

As luck would have it, the elder Issacharoff had been brainstorming new initiatives for an upcoming summit meeting between Israeli Prime Minister Shimon Peres and President Bill Clinton.

"I thought, that's not a bad idea. I went back to the ambassador at the time, who was Itamar Rabinovich, and I wasn't sure how to broach the idea," Issacharoff said. "I thought they might think I had gone a bit nuts. I said 'My kid had this idea. What do you think?' " It didn't take long before Rabinovich said "Go for it." U.S. officials cleared the idea with NASA and the National Security Council.

"President Clinton announced in a press conference after Peres had come and they had their first meeting, that they had agreed to include an Israeli astronaut on a future space shuttle mission," Issacharoff remembered later. "This is kind of a rare occasion when a diplomat and his son can actually think of something and see it become . . . a reality."

NASA and the Israel Space Agency signed a formal agreement in 1996. But neither side wanted a purely symbolic mission, a flight in which the future Israeli astronaut was merely a passenger. The Israeli mission would have to be grounded in science, one or more investigations that would be operated in orbit by the yet-to-be-named astronaut.

"At the moment we decided to have this Israeli astronaut, we decided . . . we are not sending this astronaut just to be another tourist in space," said Aby Har-Even, director of the Israel Space Agency. He recalled the flight of Sultan Salman Abdulaziz Al-Saud, a Saudi prince who flew aboard the shuttle Discovery in 1985. An Arab communications satellite was launched during the mission, but the work was done by Al-Saud's NASA crewmates. As a "payload specialist," a crew member invited to fly a specific mission who was not a full-time astronaut, Al-Saud had had no major responsibilities.

"In the past, a Saudi relative of the king flew and spent his time describing what he saw from space," Har-Even said. "We looked for an experiment that, as NASA said, could be scientifically justified."

Both sides eventually agreed on an experiment proposed by Tel

Aviv University known as the Mediterranean Israeli Dust Experiment, or MEIDEX for short.

The idea was to study how dust particles blown off the Sahara Desert and suspended in the atmosphere might affect weather patterns across northern Africa, the Mediterranean, and the Atlantic Ocean. It was known from earlier studies that particles called aerosols can counteract the global warming produced by increasing amounts of greenhouse gases. Of the aerosols blown into the atmosphere around the world, desert dust is among the most abundant.

MEIDEX was developed to study suspended Saharan dust particles from space and to map out how the aerosols move across broad regions. A secondary goal was to collect data enabling scientists to calibrate aerosol sensors on two NASA satellites already in orbit.

The Israelis had their experiment and an agreement with NASA to launch it into space. Now they had to find someone to operate it.

"I received about 25 requests from people who said 'I can be the Israeli astronaut,' " said Har-Even. "Most of them were from people who were pilots, but some were from scientists. In the Israeli air force, they have a lot of people who have passed a lot of physical and psychological training. They have an archive on everyone who is an active pilot.

"I said, maybe it is not fair to all the other people, but we don't have time or money [for an extended search]. Let's do it this way. I told the air force to find someone they thought was one of the best who was also an engineer or scientist."

It was 1997. More than a year had passed since Dean Issacharoff's innocent question in the National Air and Space Museum.

Col. Ilan Ramon, head of weapons development and acquisition for the Israeli air force, was at headquarters one evening when his phone rang. He picked it up and heard someone ask if he was interested in becoming Israel's first astronaut.

Ramon thought it was a joke. In Israel, the term "astronaut" usually refers to someone with his or her head in the clouds. Ramon had never even considered the possibility of becoming a real astronaut.

"Sometimes, when I watched movies about space and Apollo, I would envy them," Ramon said later. "But I never dreamed about going."

But it quickly became obvious the caller was serious. This was no joke. "I told him, of course I want to, but let me ask my wife first," Ramon said.

The Israel Space Agency would have been hard-pressed to find a better choice. Ramon was a veteran war fighter, a team player and a gifted pilot with 4,000 hours flying time in high-performance jets, including the state-of-the-art F-16. He was a family man with a degree in electronics and computer engineering. And his family history mirrored his nation's troubled history.

His father, a Zionist, fled Nazi Germany in 1935 after Adolf Hitler's rise to power. He settled in Tel Aviv and later fought for Israel's independence. His mother, then a teenager in Poland, refused to leave her ailing mother, and both ultimately were shipped to Auschwitz. They spent a year and a half in the infamous death camp before being liberated by Russian troops. Other relatives weren't so lucky. Ramon's mother and grandmother eventually made it to Israel in 1949.

Ramon was born in a Tel Aviv suburb in 1954, the second of two children. The family moved to Beer Sheva eight years later, and after graduating from high school in 1972, Ramon joined the air force and began compulsory military service. He served during the Yom Kippur War in 1973 and became a fighter pilot the following year, training in A-4 Skyhawks and French-built Mirages.

In 1980, Ramon joined an elite team selected to form the core of Israel's first F-16 fighter squadron. The team was sent to Hill Air Force Base near Ogden, Utah, for training. Returning to Israel a year later, Ramon was named deputy squadron commander. He played a key role in a now-legendary June 1981 bombing raid that destroyed a nuclear reactor under construction near Baghdad. One year later he took part in bombing raids intended to drive Palestinian guerrillas out of Lebanon.

Ramon never discussed his military exploits, saying simply "I would rather not talk about that." But friends and colleagues were quick to praise his skills, if not to provide details.

"He was a terrific fighter pilot," recalled Haim Ashkenazy, an executive in the aviation and aerospace department of the Israel Export Institute. "He was born to be a pilot. He had it in his blood. He was very good at it.

"He participated in a few very special operations that I can't talk about. I remember him as being part of the whole group that was chosen to perform this very, very special task.

"It didn't matter if you were air crew or ground crew, he was involved in everything and wanted to know everything—the plans, the readiness—to make sure everything went smoothly."

Ramon left the air force in 1983 to attend college, meeting his future wife, Rona, at a neighbor's party. They were married six months later, and after graduating from Tel Aviv University in 1987, he returned to the air force.

He steadily worked his way up the chain of command, serving as an F-16 squadron commander, head of the aircraft branch of the operations department, and finally head of weapons development and acquisition at air force headquarters. That's where he was in 1997 when the call came asking if he'd like to be an astronaut. One month later, the government announced that Ramon had been selected to become the first Israeli in space.

"When I was selected, I really jumped almost to space," Ramon said in a NASA interview. "I was very excited. . . . Out of several people, they selected me. So personally, I didn't do anything to get here. I think I was in the right place at the right time. That's all."

Ramon and his family prepared to move to Houston. Training at Johnson was scheduled to begin in the summer of 1998. He gradually came to understood the importance of being his country's first astronaut.

"My background is kind of a symbol of a lot of other Israelis' background," he said. "My mother is a Holocaust survivor. She was in Auschwitz. My father fought for the independence of Israel, not so long ago. I was born in Israel and I'm kind of the proof for them, and for the whole Israeli people, that whatever we fought for and we've been going through in the last century, or maybe in the last two thousand years, is becoming true."

School kids back in Israel idolized him. Scientists who were initially reluctant to spend scarce space funds on a single shuttle mission praised him for attracting children to technical subjects. Older Israelis and Holocaust survivors saw him as proof their long labors were paying off. A postage stamp was issued to commemorate his flight.

"When you talk to these people who are pretty old today, and you tell them that you're going to be in space as an Israeli astronaut, they look at you as a dream that they could have never dreamed of. So, it's very exciting for me to be able to fulfill their dream that they wouldn't dare to dream."

Like Al-Saud and other guest astronauts before him, the talented Israeli would be trained as a payload specialist. But given his scientific duties operating the MEIDEX experiment, his fighter pilot skills and the breadth of his military experience, the training teams at Johnson recognized early on that Ramon was a payload specialist in name only. His qualifications equaled those of many shuttle commanders, and there was no doubt he would be a valuable addition to some future crew.

But as 1997 gave way to a new year, no one knew when Ramon might fly.

In January 1998, far from the turmoil of the Middle East, the first component of the International Space Station was transported to the Baikonur Cosmodrome in Kazakhstan for final launch processing. The Russian-built, NASA-financed, propulsion module was designed to serve as a central supply depot and fuel dump for the long-planned orbital complex. Dubbed "Zarya" [Sunrise], the module was scheduled for launch that fall atop a Russian Proton rocket.

It was a pivotal year for NASA and its international partners after years of delays, skyrocketing costs, and congressionally mandated redesigns. It also was a pivotal year for Ilan Ramon and his future crewmates. Their eventual launch would be repeatedly delayed, in large part because of higher-priority missions to the International Space Station. MEIDEX was well and good, but the space station took priority.

Like the Apollo moon program four decades ago, the space station project had been driven by politics first and science second.

When NASA sought approval for a post-Apollo project to carry the agency beyond the old Skylab space station, agency planners envisioned development of a complex transportation and research infrastructure in low-Earth orbit. First, NASA wanted to extend the concept of Skylab and build a more ambitious, permanently manned space station that ultimately could serve as a staging base for a return to the moon or trips to Mars. A reusable orbiter, a sort of space-faring 18-wheeler, would be needed to carry the hardware into orbit and to deliver crews and supplies.

But with the war in Vietnam dominating the evening news and widespread social unrest at home, NASA had been unable to rally the political support needed for such a costly program. President Richard Nixon eventually agreed to the shuttle program, and in 1972, Congress approved funding for the Space Transportation System, or STS. But no money was available for an equally ambitious space station program. And so NASA pressed on with short-term research aboard Skylab and began building what would become the space shuttle Columbia.

NASA originally wanted to develop a completely reusable spaceship. The most ambitious schemes called for a piloted, liquid-fueled ferry craft to boost a smaller orbiter out of the dense lower atmosphere. The booster then would return to a launch site runway while the orbiter accelerated on its own into space. Flyback boosters proved too costly, however, and the spacecraft that ultimately won approval was only partially reusable.

NASA sold the space shuttle to Congress and the White House as a launch vehicle that would routinely carry huge military satellites into space, deploy commercial communications satellites, and serve as a platform for scientific research. Up to two dozen flights per year were envisioned. When Columbia blasted off on its maiden voyage April 12, 1981, construction of unmanned U.S. rockets began to be phased out.

Meanwhile, NASA planners kept pressing for a permanently manned space station, a destination for the shuttle. After the administration of President Ronald Reagan swept into office in 1980, NASA administrator James Beggs successfully lobbied to have the project inserted as a "new start" in the president's 1984 State of the Union address. The station would be developed within the decade, the president said, coinciding with the 500th anniversary of Columbus' discovery of the New World. The permanently manned outpost was expected to cost around $8 billion, not counting the price of the shuttle transportation.

That same year, shuttle Discovery joined Columbia and Challenger in NASA's shuttle fleet. Atlantis followed suit in 1985. Top-secret military missions were launched, daring satellite rescues were staged by spacewalking astronauts, and a steady stream of communications stations were deployed for paying customers around the world. Senior NASA managers and many in Congress considered the program "operational." The flight rate kept inching up. Nine missions were launched in 1985, and 15 were planned in 1986.

Then, on Jan. 28, 1986, Challenger disintegrated 73 seconds after blastoff. It was the worst disaster in space history, claiming the lives of commander Francis "Dick" Scobee, pilot Mike Smith, flight engineer Judy Resnik, Ellison Onizuka, Ron McNair, Hughes satellite engineer Greg Jarvis, and Christa McAuliffe, a New Hampshire high school social studies teacher.

The presidential commission that investigated the disaster blamed the mishap on a poorly designed joint between two of the four segments making up Challenger's right-side solid-fuel booster. The joint ruptured, the base of the booster pulled free, and the external tank was ripped open. Challenger broke up as the tank disintegrated, torn apart by aerodynamic forces.

In the weeks and months that followed, the Department of Defense abandoned the shuttle and shifted its high-priority national security payloads to unmanned Titan 4 rockets. The European consortium Arianespace, operator of the primarily French-built Ariane rocket, stepped into the vacuum and quickly dominated the free world's commercial launch market. President Reagan, in a bid to restart America's dormant unmanned-rocket industry, signed an executive order banning commercial satellites from the space shuttle.

Construction of whole fleets of unmanned rockets did, in fact, revitalize America's launch industry. The Delta and Atlas rockets built by Boeing and Lockheed Martin today are a direct result of that policy. But Reagan's executive order and the Pentagon's shift to unmanned launchers left NASA's space shuttle with just one-third of its original customer base.

The effects were not immediate. It took several years after flights resumed in 1988 to clear a backlog of high-priority interplanetary probes, science craft like the Hubble Space Telescope, and a string of research missions. But the writing was on the wall. By the mid-'90s, the space station was the shuttle's only viable customer, and from that point forward, the two programs marched in lockstep, each one dependent on the other for survival. Without the space shuttle, the station could not be built. Without the station, the shuttle had nothing to launch.

Originally dubbed "Freedom" by the Reagan administration, the now-unnamed space station ran into problems of its own in the late '80s and early '90s. NASA had vastly underestimated the original $8 billion price tag, and the agency was forced to continually scale back the program, reducing the lab's capabilities to rein in soaring costs.

The Clinton administration went a step further. In exchange for continued White House support, NASA was ordered to bring the Russians into the international project, joining Canada, Japan and the multi-nation European Space Agency. The idea was to take advantage of Russian expertise in life support systems, automated spacecraft refueling and supply ships, and experience operating their own Salyut and Mir space stations. Some believed Russian participation could save U.S. taxpayers billions of dollars.

The former Cold War adversaries quickly got to work, mounting a series of successful shuttle missions to the Mir space station to perfect the management techniques and procedures needed for assembly of the much more complex international station. The last of nine shuttle-Mir docking missions was carried out in June 1998, the month before Ramon began training in Houston.

Earlier in the shuttle program, NASA mounted microgravity research missions—flights taking scientific advantage of the weightlessness of space—using a European-built module called Spacelab. The costly laboratory module, carried aloft in the shuttle's cargo bay, was retired in the mid-1990s as NASA shifted its focus to research aboard the space station. But station assembly was repeatedly delayed. An American company, Spacehab Inc., already was providing pressurized cargo modules to NASA. With Congress concerned about delays in station science, the company mounted what insiders said was an intense lobbying campaign to drum up congressional support for additional shuttle research missions using Spacehab modules to carry experiments.

In October 1998, one month before Zarya's launch, lawmakers debating NASA's fiscal 1999 budget appropriated $15 million for "a shuttle mission which accommodates research payloads." The dollar amount was, in a sense, meaningless because a shuttle flight actually costs $400 million to $500 million, depending on how one does the math. But the appropriations language ensured that such a mission would make its way onto NASA's launch manifest.

Spacehab viewed the agreement as a win–win arrangement for the company and NASA. Critics of such space-based research, however, argued the staggering cost of a shuttle flight far outweighed any potential scientific gains. But the idea won the necessary political backing and funding language was anonymously added to NASA's fiscal 1999 budget.

Shuttle mission STS-107 was born, the last non-station research flight on NASA's books.

As Ramon was learning the ins and outs of shuttle operations, working his way through generic payload specialist training, STS-107 was assigned to Columbia. At that time, Columbia was restricted to non–space station flights because it was heavier than NASA's other orbiters and could not carry large components to the station's orbit. As a result, NASA's few non-station payloads had been put on Columbia: an X-ray telescope known as the Chandra X-ray Observatory; two missions to service the Hubble Space Telescope; and now, STS-107.

Because Columbia's mission was the final nonstation research flight left on NASA's books, "everybody wanted a piece of it," said Mike Hawes, a senior station planner at NASA headquarters. "There were a lot of competing interests."

As 1998 drew to a close, the question was what—and who—would fly on it?

At the beginning of that same year, Vice President Al Gore had a dream. He dreamed of a small spacecraft that would beam live color pictures of Earth back to the Internet where anyone, anywhere in the world, could marvel at the home world spinning in the deep black of space. He believed such images would be a stimulus for environmental research, inspire students around the world to pursue careers in math and science, and provide valuable scientific data. The satellite was dubbed "Triana," named after the lookout on one of Columbus's ships who first spotted the New World in 1492. The spacecraft was expected to cost between $25 million and $50 million.

Gore was heavily favored to be the Democratic presidential candidate in the 2000 election, and Republicans wasted little time attacking Triana. They promptly labeled the project "Gore-sat" or "Gore-cam," or they simply derided it as "junk science." To then NASA administrator Dan Goldin, however, a vice president's dream, no matter how outlandish many considered it, could not be ignored. By the end of the year, a contract was awarded for assembly of the satellite's main camera, and launch was targeted for November 2000 aboard Columbia as part of mission STS-107.

Triana was a small part of Columbia's overall payload. NASA "emptied the closet" to load the mission with space station–class research in a variety of disciplines, ranging from biology to medicine, from materials science to pure physics and technology development. More than 80 experiments ultimately were assigned to the flight. The bulk of the research would be carried out inside a roomy, double-wide Spacehab module mounted in Columbia's cargo bay and connected to the ship's crew cabin by a pressurized tunnel.

Columbia would carry enough supplies to remain in orbit 16 days, a long-duration mission by shuttle standards. To collect as much data as possible, the crew would work around the clock in two 12-hour shifts.

But STS-107's wait was far from over. Next up for Columbia was launch of the Chandra X-ray Observatory, which would slip nearly a year because of technical problems with the telescope and its attached booster. When the shuttle finally did take off in July 1999, wiring problems triggered a frightening short circuit requiring fleetwide inspections and repairs. Columbia went into the shop for an already planned overhaul under a cloud of uncertainty about when it might return to flight.

In the meantime, an internal NASA audit raised questions about Triana's political roots. Lawmakers, concerned that the satellite's launch could give Gore an election-season boost, ordered a work stoppage until an independent assessment by the National Academy of Sciences could be completed. By the time the NAS issued its report concluding the satellite did, in fact, have scientific merit, it was too late to complete preparations in time for launch aboard Columbia. The satellite was bumped from the mission in July 2000.

NASA began looking for other experiments that could take advantage of unused space in Columbia's cargo bay. Israel had a good candidate to help fill the void—MEIDEX—and the man to operate it: Ilan Ramon. NASA agreed, and on Sept. 28, 2000, the agency issued a press release naming Ramon and four NASA astronauts to the STS-107 mission. The announcement was typically brief:

> Four mission specialists and one payload specialist have been assigned to the STS-107 mission, undertaking a series of U.S., international and commercial experiments. Michael P. Anderson and Kalpana Chawla will be joined by Mission Specialists David M. Brown and Laurel B. Clark, both members of the astronaut class of 1996 and first-time fliers. Payload Specialist Ilan Ramon will round out the crew. Anderson and Chawla both have one previous flight to their credit, STS-89 in 1998 and STS-87 in 1997, respectively. A commander and pilot will be named at a later date.

A little more than two months later, NASA announced that Rick Husband, making his second flight, would command STS-107. Rookie Willie McCool would serve as pilot. At that point, the launch was targeted for July 2001. But the Chandra delays, the wiring repairs, Columbia's routine maintenance, and the first of two planned Hubble servicing missions would conspire to delay STS-107's launch another year, to July 19, 2002.

Husband's crew had no complaints. They were in the launch pipeline, and sooner or later, they would be flying in space. They had plenty of time to train and get to know one another.

It takes a space shuttle about 36 seconds to climb 13,000 feet. But on foot in remote Wyoming in August 2001, it took Husband and his six new crewmates six hours to reach that height, at the top of Wind River Peak. The exhausted astronauts put a small cloth shuttle patch on the summit and posed for a group photo. Clark, smiling broadly, her hair windblown, sits legs akimbo atop a granite boulder, her right arm draped over Brown's shoulder. Husband stands off to her left, one arm around Ramon's shoulder, the other on McCool's. Chawla smiles serenely in the foreground as Anderson, wearing a backward baseball cap, displays the shuttle patch with a satisfied grin.

One of their two guides whipped out a heretofore hidden cell phone and dialed friends back at Johnson. The astronauts called out in unison, one yelling "hey!" and another "13,000 feet!" After a few moments of silence, looking out across the rugged Wind River Mountains terrain, Husband began singing. The others joined in and the familiar words of the Doxology—"praise God, from whom all blessings flow"—drifted into the clear Wyoming sky.

The climb capped a 10-day stint in the Rocky Mountains wilderness organized by the National Outdoor Leadership School. NASA had begun sending astronauts to the school two years earlier to forge leadership skills with a challenging test of physical endurance and self-reliance. The astronauts had no maps, CD players, or cell phones. They took six-mile hikes each day wearing heavy backpacks. They took turns being the leader. And then they climbed Wind River Peak as a final test, deciding to make the ascent together, as a team, rather than split up. It was an exhilarating experience, an achievement that created a bond of friendship that went beyond the usual camaraderie most crews experience.

"We all went out into the mountains and we backpacked with 50-, 60-pound packs for nine nights and 10 days out there," Husband recalled later. "We got to learn a lot about how each of us, as individuals, deal with the kinds of situations they put us into. It's a physical challenge . . . and then learning to work together, pulling together and learning more about each other.

"We all, I think, understand each other probably as well as a crew does when they come back from a space mission."

"They bonded so well," recalled Darla Racz, astronaut training manager at Johnson. "They actually loved each other like family."

Ramon, Brown, Anderson, Chawla and Clark began training for their mission the month before their names were announced to the public. Husband and McCool joined the following October. They would spend the next 900 days together, working through delay after delay and getting to know each other, their families and their life histories in a way few other crews ever manage.

"This crew was made up of truly exceptional people," said

McCool's wife, Lani. "Not one of them had an ego and because of this, the respect they had for each other was genuine, never competitive."

Ramon, 48, fit right in. "Life is not a rehearsal," he liked to say. The crew trainers had especially fond memories. During a long simulation near Christmas, trainer Stephanie Turner left her instructor's station to locate another headset.

"I was coming around the corner and I looked left and there was Ilan standing there in his blue flight suit underwear and a Santa Claus hat," she said. "And it's not hanging down or anything, it's pointed straight up in the air. He had a, you know, 'I'm 14 and I'm waiting for Santa on Christmas morning' kind of look to him. And he's got this grin. And Willie and Dave were also down there and they're just kind of looking at me and my face is turning red because you know, technically, it's their underwear, you know?"

Husband set the tone. A family man with a ready "Howdy" and firm handshake for friends and strangers alike, he was born and raised in Amarillo, Texas. Husband loved his hometown so much that early in his NASA career he was dubbed "BFA"—Boy From Amarillo. He never forgot the fact that many of Amarillo's younger generation followed his exploits in space.

"It's nice to see the hometown folks get so excited about this and see a lot of attention generated there," Husband said before his first launch. "From the standpoint of all the kids in school, I think it's good for them to see some coverage and get more involved in the mission."

Lots of other astronauts dutifully said the same sorts of things and also made the rounds at schools, ever mindful that deft public relations skills never hurt anyone looking to move up the NASA ladder and get flight assignments. With Husband, however, no one ever doubted he meant every word.

"I think Rick's unique ability was to bring people together and get them to work as a team," said astronaut Bob Cabana. "He had a real and innate ability to see what people could do and to get them to work together and direct them without being dictatorial or telling folks precisely what to do. He is one of those guys that could tell you what to do and make you feel like it was your idea."

Husband's mother, Jane, still lived in Amarillo in the same house where he grew up. His father, Doug, owner of an Amarillo meatpacking house, had died a decade earlier. On this frequent trips home, the shuttle commander would mow grass, rake leaves and shovel snow, not only for his mother, but for her 74-year-old next-door neighbor.

"Somebody remarked to me that this guy almost seems perfect, almost too good to be true," said Ann Micklos, Brown's friend. "I mean he is a God-fearing, all-American boy. This guy is a shuttle commander and he is going home on the weekends to mow his 80-year-old mother's lawn and the next door neighbor's lawn."

Husband still owned his first car, a mint-condition 1975 Camaro, which he lovingly maintained and drove to work at the Johnson Space Center every day.

"Rick was a cool drink of water, he was a long, tall Texan from Amarillo," remembered Lisa Reed, a former astronaut trainer at Johnson who worked with Husband from the beginning of his career. "Probably one of the nicest people you'd ever meet. He always had a smile and the most beautiful singing voice that you've ever heard. A very deep baritone and he sang beautifully. He loved George Strait and the rodeo."

Astronauts are stars at Johnson and the other manned spaceflight centers. They are the ones who put their lives on the line, the inheritors of the "right stuff" mantle of the Mercury, Gemini and Apollo astronauts. Some astronauts revel in that macho image, but Husband appeared to avoid it. He went out of his way to make personal contact with lower-level engineers and technicians, a trait that endeared him to his trainers.

One day, as Husband and his crewmates were climbing into one of the shuttle simulators at Johnson, preparing for yet another session, Turner called out "Good morning, sir." A routine greeting before a normal training run.

"And he's like, 'Hi, Stephanie.' And he gets down to the next door and he's got his hand on the doorknob, and he turns around and comes back and he puts his hand on my shoulder and says, 'I just wanted to tell you you're doing a great job.' And that was just it for me. I would have moved the Earth for him. He was just a motivator."

McCool, a slender, boyish-looking Navy commander with an engaging grin, shared Husband's humble demeanor almost to an extreme.

An Eagle Scout born in San Diego, the 41-year-old pilot spent part of his youth growing up in west Texas and earning a reputation as one of Lubbock's best long-distance runners. He graduated second in a class of 1,083 at the U.S. Naval Academy in 1983 and later earned master's degrees in computer science and aeronautical engineering.

"He is just humble enough that he probably could have graduated No. 1 and intentionally didn't, just so that he didn't get all the attention," astronaut Kent Rominger joked. "I wouldn't put that past him. That is the kind of guy he was. He would probably have said, 'You know, that would mean a lot more to old Joe to be the No. 1 guy than me and I don't care about the attention.' "

McCool earned his wings in 1986 and completed two tours aboard the aircraft carrier USS Coral Sea, flying Prowler electronic warfare jets before being selected to attend Naval Test Pilot School at Patuxent River, Maryland. After graduating, he remained as a test pilot in the Strike Aircraft Test Directorate, concentrating on flight testing an advanced version of the Prowler.

He then returned to his original base at Whidbey Island, Wash., and was assigned to the USS Enterprise. By the time NASA selected him as an astronaut in 1996, he had logged 400 carrier landings and nearly 3,000 hours flying in high-performance jets.

"Willie was an incredibly passionate person and everything he experienced, whether it was running through trails in forestlands, flying at low level or reading a bedtime story to one of our boys, he never took anything in life for granted," Lani said. Launching aboard Columbia "was no different. He saw it as the 'ultimate experience' in his career. He promised to remember every second of the experience to share with us later on."

McCool had known Lani, a native of Guam, when he was living there as a teenager. They moved apart, but years later he tracked her down. They eventually got married and had three sons, Sean, Christopher and Cameron.

"I never saw Willie down," Cabana said. "I never saw him say anything bad about anybody. He was really soft-spoken. You could ask him to do something, and it was done perfectly the next day."

Lani described Willie with a quote from a former astronaut: "He had the heart of a child, the brain of a genius and the courage of a lion."

When a crew first comes together for training, the commander faces a critical decision: Who among them would make the most effective flight engineer, a position known as "mission specialist 2." During launch and re-entry, the flight engineer, sitting directly behind and between the commander and pilot on the shuttle's upper flight deck, serves as a second set of eyes, making sure the two pilots don't miss anything. The flight engineer must be an expert on the shuttle's complex systems and be ready to respond instantly if something goes wrong.

For Columbia's crew, Husband had an obviously good choice: physician-astronaut Dave Brown. Even in the world of super-achieving astronauts, Brown stood out: a Navy flight surgeon and aviator who graduated first in his class in pilot training. Brown had 1,700 hours of flying time in jet warplanes and was the only flight surgeon in a decade to be selected for pilot training.

But Husband named Chawla his flight engineer, a choice that both raised eyebrows and won praise.

With a doctorate in aerospace engineering, Chawla, known in the astronaut office by her initials, "KC," was an accomplished pilot in her own right, holding a flight instructor's license and a commercial pilot's license in single- and multi-engine aircraft. The 41-year-old engineer enjoyed acrobatics in open-cockpit biplanes and frequently took friends aloft for dizzying spins above Texas. Reed recalled meeting her at a local airport after work one day.

"She walked me over to a big cabinet in the hangar and threw the doors open and she goes 'Pick one.' I looked down and they were parachutes. I said 'Well KC, what kind of flying are we going to be doing?' Because normally you don't want to bail out of a perfectly good airplane. She was doing aerobatics. We went through just about every possible maneuver you could think of."

Her interest in flying dated back to childhood in Karnal, India, a small town that had an active airport and flying club. Riding bikes with her brother, she became fascinated watching the small planes zooming by overhead, wondering where they were going and what it would be like to fly.

"Every once in a while, we'd ask my dad if we could get a ride in one of these planes," she said in a NASA interview. "And he did take us to the flying club and get us a ride. . . . I think that's really my [earliest] link to aerospace engineering."

By the time she was in the eighth grade, the point in the Indian school system when a student picks a general academic theme to pursue, Chawla knew what she wanted to do with her life—aerospace engineering—and began concentrating on physics, chemistry, and mathematics. After earning a degree from Punjab Engineering College in 1982, she headed for the United States to pursue advanced degrees and broader horizons. Two years later, she met her future husband at the University of Texas at Arlington.

She had no plans to be an astronaut, an idea she considered "really, really farfetched." But in 1988, Chawla went to work at NASA's Ames Research Center near San Francisco, earning a reputation as a top-notch researcher before moving into the private sector as a senior research scientist with an expertise in aerodynamics.

In 1994, NASA accepted her application to become an astronaut, and she began training with the space agency's fifteenth astronaut class, the same group that included future crewmates Husband and Anderson. She quickly earned a seat on the shuttle, blasting off aboard Columbia in 1997. What happened on that mission would dog her career.

Chawla was responsible for using the shuttle's 50-foot robot arm to recapture a small sun-study satellite that failed to properly activate after its release. During the retrieval operation, the arm appeared to nudge the satellite ever so slightly, causing it to begin a slow tumble. Despite repeated attempts, Chawla was unable to snag the spacecraft, and the shuttle ultimately had to back away. Two astronauts carrying out a spacewalk later in the mission saved the day by manually plucking the satellite out of open space and re-stowing it in the cargo bay.

NASA managers later hinted that operator error played a role in the missed grapple, and reporters asked painfully pointed questions during an orbital news conference.

"I didn't know everybody was pointing fingers at me until now," Chawla replied to one question with a laugh. They were. When she returned to Earth and re-entered the general astronaut pool awaiting future assignment, some observers wondered if she would ever fly again.

"It was really a crime that it kind of all fell out from people who knew nothing about it and watched it and said 'Hey, that was all KC's fault,' " Rominger said. "That was wrong. She should have been put right back in there. Did she make a mistake with the arm? She did. But, all these other experienced guys in the office could have missed grapples and messed them up, too."

Dave Pitre, a member of the STS-107 training team, said he believed Husband was "concerned about her future and thought, by selecting her as [flight engineer], it would give her a chance to re-prove herself, she could regain her confidence and the astronaut office would see that she was a viable selection in the future."

Regardless of her motivation, Chawla mastered every aspect of the flight engineer position, impressing trainers and fellow astronauts alike with the depth of her knowledge, her endless hours of study and her phenomenal memory. Brown would joke with her, "KC, can I borrow your brain?"

Anderson also could have served as flight engineer, but he was put in charge of Columbia's many payloads, responsible for overseeing the integration of the scores of experiments added to the shuttle's mission. His official title was Payload Commander.

"Well, exactly what does that mean?" Anderson asked early on. "It didn't take me long to find out. When you have a complex mission like this, you need a point of contact for answering all the crew questions. A person to go to, to help make the decisions about things and try to find the best way to make this mission a success. . . . So my job, basically, was to try to pull this mission together. Try to make sure the crew was well trained. Try to make sure that payloads were well integrated into the space shuttle."

Anderson was born in upstate New York on Christmas Day in 1959, but considered Spokane, Washington, to be his hometown. He was an Air Force brat for whom the constant roar of jets coming and going provided endless fascination. He also had a strong interest in science and had followed the Apollo moon program with excitement.

"To me, [being an astronaut] was just the best combination of the two," he said. "Here you have these men that are scientists, engineers, and they're also flying these wonderful airplanes and these great spaceships, and they're going places. And to me, that just seemed like the perfect mix and the perfect job."

Anderson, 43, went to the University of Washington and studied physics and astronomy. He went through the ROTC program to help pay for his education and was commissioned as a second lieutenant in the Air Force.

He eventually made his way into pilot training and ended up at the controls of the Strategic Air Command's "Looking Glass" airborne command post. He then served as an air refueling instructor in KC-135 tanker jets, ultimately logging more than 3,000 hours of flying time in various models.

Along the way, to broaden his educational background and to beef up his résumé, he went back to college and earned a master's degree in physics from Creighton University in 1990. He was a lieutenant colonel in the Air Force.

"At that point, after having a master's degree and a couple of thousand hours flying aircraft, I thought 'Well, if I'm ever going to make my move towards NASA, I'd better do it now.' "

NASA accepted Anderson's application to the astronaut corps in December 1994, and he joined Husband and Chawla for initial shuttle training. Like his commander and pilot, "he never bragged," Cabana said. "You just always enjoyed being around Mike and you could give him something to do, it would get done. Just really a nice guy."

He was devoted to his two daughters, Sydney and Kaycee, and his wife, Sandra. They attended the same church as Husband's family, sharing Rick and Evelyn's deep faith. Anderson looked forward to flying with his friend.

"When we first started training for this flight and we got our first briefings about the number of payloads and experiments that were going to be on this flight, I was really amazed," Anderson said. "It was certainly evident that the scientific community had been waiting a long time for this flight."

But first, the crew had to reach orbit. Anderson described launch aboard a shuttle as "a terrible way to get into space." But with no alternatives available, he said, "I'll take that ride."

"You're really taking an explosion and trying to control it," he said. "Even though we've gone to great pains to make it as safe as we can, there's always the potential for something going wrong."

But he didn't worry so much about re-entry.

"Entries are a little bit better than launch," Anderson said. "It's a little quieter, it's not quite as violent and you can enjoy it a little bit. But still for me on this flight entry, I'm just going to sit down in my seat and, hopefully, reflect on the 16 days on orbit that we've had, just anxious to get back to Earth and give the scientists all their research results. And you know, I'll be happy to have the flight behind us."

Like Husband, McCool, Anderson and Ramon, Navy Cmdr. Clark put family ahead of her career. Despite the rigors of astronaut training and the long hours away from home, the 41-year-old physician always managed to find time for her husband, Jon, and their son Iain, telling him "my most important job is being his mother."

"She was really a neat person," said Jon, a flight surgeon for the Navy, and later, NASA. "She was a multitude of people. She could be a very high-risk, technical, focused astronaut or she could be a tender loving mother, or wife, or sister, or daughter. She was very good at adapting to the situation. She was just one of those people who was able to fit in and be gracious, and a mom, in the mommy's club, taking the kids to the pool, or being part of a very intense science mission in space."

Trainer Gail Barnett knew exactly what he meant.

"I knew she was the same Laurel who was also dropping her child off at Montessori school at the same time I was, and we were always the same, both shot out of a cannon trying to get our kids to school and dash back to work and everything," Barnett said. "And the first day [preparing for a simulator run], she was trying to get something out of her purse and all these little-boy toys kept spilling out of it and I said, 'I have a purse that looks just like that!'

"She was just such a devoted mother. She would start talking about Iain and her face would just soften and her eyes would light up and we would talk about our children most of the time. We would talk about work, sure, but we spent a lot of time talking about my sons and her son."

Clark was a native Iowan who considered Racine, Wisconsin, to be her hometown. She grew up with a love of the outdoors and eventually took up such daring hobbies as scuba diving, flying and skydiving. She earned an undergraduate degree in zoology at the University of Wisconsin at Madison in 1983 and a doctorate in medicine there in 1987.

Following medical school, Clark studied pediatric medicine at the Naval Hospital in Bethesda, Maryland, and completed Navy courses in undersea medical and submarine diving officer training. She met Jon during dive classes in Florida in 1989. Her future husband got a kick out of turning her diving knife backwards when she wasn't looking. The instructor would promptly throw her gear into the pool and make her fish it out.

"She never did figure it out," Jon said, laughing.

The two parted when the classes ended. They stayed in touch and went on a few dive trips together. Clark, in the meantime, became a radiation health officer and undersea medical officer and was transferred to Scotland, where she dove with Navy divers, including SEAL commandos, and carried out medical evacuations from submarines.

"I went over and visited her quite a bit and then just over time it just slowly evolved into a more significant relationship," Jon Clark said. The two were married in 1991. The next year, they moved to Arizona. Shortly after participating in the shuttle disaster drill in Florida, Clark applied to become an astronaut.

"So she comes down for the interview, she's really pregnant and I'm thinking oh yeah, like this is going to be a real plus," said Jon. "It was pretty obvious to me they weren't going to take her because she was pregnant. But she did well and they invited her back for the next cycle, which was two years later."

She made it, and in 1996 began training to become on astronaut. Four years later, she was named to Columbia's crew, and her life became "a little bit of a whirlwind."

"And it's getting busier as we get closer to flight," she said. "I'm expecting to have a lot of fun. I'm expecting to be tired at the end. But I'm expecting it to be the experience of my lifetime so far."

Clark knew Dave Brown from their Navy days, and they were close friends. Micklos said Clark was like a big sister to him "and they were just like two peas in a pod when it came to bickering at each other and keeping each other in line."

Brown was the only single member of the crew, but his aging Labrador retriever, Duggans, was a constant companion. Brown, 46, was born in Arlington, Virginia, and was bitten by the flying bug at age 7. A family friend took him up in a small plane, and Brown always remembered the moment the landing gear tires left the ground. "It sure set a bit in my head that's been there ever since," he said years later.

Like Anderson, Brown was equally fascinated by science and technical subjects. But he never imagined he could grow up to be an astronaut.

"I absolutely couldn't identify with the people who were astronauts," he said. "I thought they were movie stars and I thought I was kind of a normal kid. So I couldn't see a path, how a normal kid could ever get to be one of those people. It was really kind of much later in life after I'd been to medical school, I'd gone on to become a Navy pilot, that I thought, 'Well maybe I have some skills and background that NASA might be interested in.' "

Brown earned a degree in biology from the College of William and Mary in 1978, lettering in gymnastics and working part-time with Circus Kingdom doing acrobatics, riding a seven-foot unicycle, and walking on stilts. He earned a doctorate in medicine in 1982 from the Eastern Virginia Medical School.

He joined the Navy after his medical internship in South Carolina and served in Alaska and the western Pacific before applying for pilot training. The first application was rejected, but he persisted, and the second time around, he was accepted. He became a naval aviator in 1990, finishing first in his training class and qualifying for duty as a carrier pilot. He later flew F-18 Hornets and became a flight surgeon at the Naval Test Pilot School.

Brown was named an astronaut in April 1996, and when he was selected for the crew of STS-107, no one was surprised. He was a perfect fit.

"He was not really one to talk about himself, he would always put the spotlight on someone else and let them do the talking," Micklos said. "He was very humble."

Reed recalled Brown's help and comfort when her own dog developed cancer. She had mentioned her dog's illness to him in passing one Friday afternoon. Two days later, he called and asked if he could come over.

"He showed up at my house, and he had a stack of papers in his hand. He had done research on this type of tumor my dog had, all treatments, and he sat down with me and explained from a medical standpoint how to make a better decision on what to do. The choices were to put her to sleep or put her through a really horrible surgery. He spent a lot of time with me. I said 'Dave, how long did this take?' He said 'I started Friday night and I just got finished this morning.' I just thought what a great person to spend this time because he knew it was important to me."

Brown was an avid amateur videographer, and he carried a camera everywhere he went, documenting the minutiae of training as the crew prepared for flight. "Dave would always come to the instructor's station with his camera," said trainer Lisa Anderson. "He would put it into your face and you were supposed to act natural. Finally he came up with 'Just act like a fuzzy brown squirrel.' And that became kind of our mantra. Whenever he came in with a camera, we were all 'fuzzy brown squirrels, everyone!' "

The saying became an ongoing gag and Anderson brought in a Kung Fu hamster toy. "I thought well, I know he's not a squirrel but he kind of looks like one." The crew quickly adopted the toy as a sort of running joke. "He really became a good mascot for us," Anderson said.

Brown was outwardly enthusiastic and as gung ho as the rest of the crew, but he told at least one trainer and another friend that he wasn't sure he would make it back to Earth.

"He had a premonition he was going to die, to burn up in space," Racz said. "He stepped up behind my desk one day and he said 'I had a dream.' It was a dream, a vivid dream to him."

Another friend later said, "He told me he was going to die on this mission. I kid you not. May 2001. It shook me to my core. He looked at me and he said, 'You know, I don't know if I want to stay around at 40-something. Once I get this flight under my belt, I may leave. But you know, I just know I'm going to die on this mission.' I'm like, 'Dave, nobody wants to hear something like that. It gives you the willies. He looked me square in the eye, and I didn't doubt him. I didn't ask him why. I didn't want to know."

Training reached fever pitch in the spring of 2002. The launch was scheduled for July 19. But in mid-June, just a few days before Columbia was to be hauled to the launchpad, STS-107 was delayed yet again.

During routine inspections of Atlantis, engineers discovered a small crack inside the 12-inch-wide liquid hydrogen fuel line leading to one of the shuttle's main engines. Two more cracks were found the next day. Similar cracks then were spotted aboard Discovery and then Columbia, at which point work on Columbia's flight was suspended. Cracks then were found aboard Endeavour, and the entire shuttle fleet was grounded.

Engineers were concerned that if a crack worsened in flight and a piece of debris broke off, it could get sucked into a main engine at high velocity, triggering a catastrophe. They finally figured out a way to smooth out the cracks, but it was too late for Columbia to fly as scheduled. To keep station assembly on track, NASA managers decided to move Columbia's mission into early 2003, after a pair of higher-priority station flights. Columbia's new launch date was Jan. 16, 2003.

By now, postponements were nothing new to Husband and his crewmates. They took advantage of the delay to travel and spend holiday time with their families. All the while, they kept their fingers crossed. If the two flights ahead of them could just get off the ground, their long wait finally would be over.

THREE

"SAFE TO FLY WITH NO NEW CONCERNS"

There's a Thomas Jefferson quote about eternal vigilance being the price of liberty. Well, the same thing goes for safety. Eternal vigilance is the price that we pay to fly safely. . . . I guess the message is that as we're focusing on the trees up in front of us, don't miss the forest—what the hardware is trying to tell us and what our processes are trying to tell us.

—Bill Readdy, NASA's associate administrator for the Office of Space Flight, at the Flight Readiness Review for the shuttle Endeavour's November 2002 launch.

Gasps filled Kennedy's Image Analysis Facility as film reviewers screened what looked like the largest debris strike ever to hit a shuttle.

Seconds after liftoff, a chunk of foam insulation broke

free from the 15-story external fuel tank that holds more than a half-million gallons of liquid oxygen and liquid hydrogen rocket fuel. The mailbox-sized chunk tumbled down along the tank and hit one of the shuttle's twin rocket boosters an instant later. Long-range tracking cameras around Cape Canaveral captured stunning images of the event as the shuttle hurtled skyward. Hours later, jaws would drop again at Johnson and the Marshall Space Flight Center in Huntsville, Alabama, when other NASA film review teams got their first look at Atlantis' October 2002 launch.

A dozen or so people stared at the photos on Oct. 8, the day after liftoff, in the film laboratory—a large, brown, windowless room on the third floor of Kennedy's Vehicle Assembly Building, the mammoth facility where the orbiter, fuel tank and boosters are put to-gether for flight. Engineers spread out around five film-cluttered conference tables to debate the strike's origin as it was projected on a giant screen that covered an entire wall of the room. Occasionally, someone would walk over to a 4-foot model of the shuttle that stood in front of the screen to present his theory on what part of the tank the foam had fallen from. There was no way to be sure with the limited camera coverage. However, the images clearly showed where the foam had hit: on a metal ring covered with insulation that attached the base of the left-side booster to the fuel tank.

"We didn't have a good view of the forward end of the shuttle tank," said Armando Oliu, a veteran NASA film reviewer who headed Kennedy's ice, debris, and final inspection team. "So all we saw was the debris coming down the external tank and then impacting."

On the booster near the attachment ring was a critical electronics box that relayed computer commands from Atlantis. The box transmitted signals that swiveled the booster's nozzle to steer the shuttle during the first stage of flight and jettisoned the booster after its fuel was spent. Had the box been disabled by the impact, the result could have been catastrophic. However, the booster had performed as expected. Explosive charges fired as planned two minutes and four seconds after launch to drop it into the Atlantic Ocean. Atlantis was safely in orbit. The foam must have missed the electronics box. There would be no need to convene a team of engineers to assess possible damage to the ship.

"We knew that there was no problem with the booster's performance," Oliu said, "and we knew that the debris did not strike the orbiter."

Even so, Bob Page, one of Oliu's colleagues, was deeply concerned. Page had a reputation at NASA as a conscientious, no-nonsense manager with an eagle eye for spotting unseen problems in film reviews and an almost religious devotion to shuttle safety. The 42-year-old electrical engineer had grown up in the shadow of the space program. His father was an engineer at Marshall who was sent to Kennedy in 1964 to help design ground support equipment for the giant Saturn moon rockets.

As a boy, Page had watched the launch of Apollo 11 from his backyard and built model rockets to emulate the exploits of his heroes. After graduating from a local high school and college, he went to work at a McDonnell Douglas plant that built cruise missiles in Titusville, Florida, just across the Indian River from Kennedy. He saw Challenger explode from a second-floor window there and applied for a job at NASA several months later with the hope of doing his part to help the shuttle return to flight. Page was initially hired by Kennedy's vehicle engineering department in 1987 as a go-between with the shuttle's spare parts depot in the nearby city of Cape Canaveral. He began analyzing film and in 1993, was named chairman of the Intercenter Photo Working Group, an agencywide organization responsible for film coverage of shuttle launches, missions, and landings.

Page immediately understood the implications of a large debris strike on that part of the booster. Only providence had prevented disaster.

"If we had lost the electronics to the booster during ascent," Page said, "it would have been a bad day."

Foam falling off the external tank and striking the shuttle was nothing new to Page, Oliu and other engineers. It happened during every launch. Their concern, however, usually focused not on the boosters, but on the fragile heat shielding that protects the orbiter from extreme temperatures during the fiery plunge home through Earth's atmosphere. Out of 77 previous shuttle launches from which photography was available, cameras had documented foam loss on 63. On all of the remaining launches—night liftoffs and cases where cameras failed to capture shedding or were inconclusive—unseen impacts were known to have occurred because of divots carved in some of the 24,000 protective heat tiles that cover the orbiter. Made of silica fibers derived from common sand, the tiles were easily damaged.

A shower of foam dinged more than 300 tiles when Columbia made the first shuttle flight in April 1981. During the 21 years since, orbiters had returned after each mission with an average of 143 damaged tiles, including 31 with scars measuring more than an inch long. An average of 101 of those divots were on the belly and lower surfaces, where especially critical black-colored tiles are exposed to temperatures of up to 2,300 degrees during reentry. The nicks seldom penetrated more than an inch into the black tiles, which typically measure 6-inches square and range from 1 to 3 inches thick. But occasionally, the gouges would cut into or through a layer of denser material at the tiles' base that was a last line of defense for the ship's vulnerable aluminum airframe underneath. It was a major concern. Damage was far less common, on the other hand, to another key part of the orbiter's thermal protection system: the denser, reinforced carbon-carbon (RCC) composite material used to protect the leading edges of the ship's wings and its nose from the most extreme heating.

Foam had hit the shuttle during every one of the 110 missions that preceded Atlantis' Oct. 7 liftoff. That reality flew in the face of a clear-cut rule against debris strikes that had existed since the dawn of the program. Volume 10, Book 1, of the National Space Transportation System 07700 document, nicknamed "the shuttle bible" by NASA engineers, laid out the flight and ground system requirements in no uncertain terms:

> The Space Shuttle System, including the ground systems, shall be designed to preclude the shedding of ice and/or other debris from the Shuttle elements during prelaunch and flight operations that would jeopardize the flight crew, vehicle, mission success, or would adversely impact turnaround operations.

The rule never had been enforced. Shuttle program managers decided in 1988 that foam and other materials coming off the external tank and striking the orbiter was an "accepted risk." A study with the arcane name of *Integration Hazard Report* 37, one of dozens of constantly updated reports in a NASA database of potential threats, determined there was a "remote" chance that falling tank debris could cause catastrophic damage. Potential effects noted in the report included damage to the orbiter's thermal protection, structural overheating and "loss of vehicle, mission and crew."

Getting rid of the foam wasn't an option. It was an essential part of the external tank, providing an outer layer of thermal insulation 1 to 2 inches thick in most spots. The insulation was developed to help keep the tank's propellants chilled to temperatures as low as 423 degrees below zero in the central Florida climate while preventing ice from building up on the tank's exterior.

NASA covered the tanks with four different types of the polyurethane-based foam, all having the same basic weight, look and feel as their cousin, polystyrene, commonly used to make Styrofoam. Depending on the location on the tank, computer-controlled machines or workers using spray guns applied the foam at the 832-acre Michoud Assembly Facility in east New Orleans, a plant operated by aerospace contractor Lockheed Martin under management from NASA's External Tank Project Office at Marshall. Technicians at Kennedy also added foam by hand or with a sprayer to a few final "closeout" areas after the tanks were connected to the orbiter and the boosters.

When there were unusual cases of major foam loss, dubbed "out of family" in NASA parlance, the agency usually took steps to fix them. During the Nov. 19, 1997, launch of Columbia—Chawla's first flight—foam fell from two panels on the tank's ribbed midsection and caused 308 dings, more than double the average number. The phenomenon, nicknamed "popcorning," was attributed to the removal of Freon from the foam's application process in an effort to make it more environmentally friendly.

To solve the problem, engineers reduced the foam's thickness in the spots where it fell off, poked holes in the area to allow trapped gases beneath to escape, and began sanding off the hard outer crust. Meanwhile, the shuttle continued to fly while the fixes were being developed. NASA managers concluded ten missions later that shedding had been reduced to an "acceptable" level again, and the issue was closed.

One of the toughest spots on the tank for Michoud workers to manufacture with precision and consistency was the so-called bipod area at the bottom of the tank's midsection. There, a pair of foam ramps covered two metal spindles. Connected to the spindles were struts that attached the tank to a spot beneath the orbiter's nose. The suitcase-sized ramps had several purposes: They provided a streamlined surface that protected the spindles and related hardware from aerodynamic forces during launch; they safeguarded the spot from heating during the climb to orbit; and they prevented the buildup of ice before liftoff. Workers built the ramps by spraying layers of foam on the spindles, then trimming and sanding the lumps down to the proper shape and size. Despite their best efforts, the process occasionally would leave small undetected voids between the ramp and a coating underneath that covered the tank's aluminum-alloy skin.

Never had a case been documented where foam had fallen off the bipod ramp on the tank's right side. Tank engineers speculated that a nearby 17-inch wide liquid oxygen line helped shield the right ramp from stresses during launch. But the left ramp had shed debris at least three times that shuttle managers knew about before the Atlantis launch, and some of the largest foam chunks ever seen to fall off the tank had come from there. A 19- by 12-inch piece of the left ramp came loose during a June 1983 launch of Challenger. Engineers never figured out the cause, but several missions later, the ramp's design was changed from a 45-degree angle to an aerodynamically sleeker angle of 30 degrees or less. A second incident occurred in January 1990, when a large piece of foam, including part of the left ramp, fell off during the launch of Columbia. Engineers blamed the shedding on air trapped in voids between the foam and the material underneath it. The solution was to poke small vent holes in the foam around the bipod area. The third and most recent incident took place in June 1992, when a huge piece of left ramp foam, estimated to be 26 by 10 inches, broke free during launch and again hit Columbia. After landing, the damage area was the largest ever seen on the shuttle's tiles. Shuttle managers again blamed the incident on poor venting and kept drilling holes.

The acceptable size of foam strikes expanded over the years as shedding continued. Incidents increasingly came to be seen as "in family"—that is, previously seen, analyzed, and therefore accepted—as the orbiters continued to survive impacts. Tank engineers became increasingly confident they understood the issue and took comfort in the fact that efforts were being made to eliminate the worst cases. No one gave much thought to the hypothetical "loss of vehicle, mission and crew" scenario mentioned in *Integrated Hazard Report* 37.

Intellectually, shuttle managers knew it was possible to fatally damage the tiles and reinforced carbon-carbon surfaces with big foam strikes, and that large foam pieces could fall from the bipod ramp. But all of their experience indicated otherwise. There were only three documented cases where large pieces of foam had fallen from the left ramp. There was no evidence any had been shed from the right. Studies showed the likelihood of a ramp breaking off and hitting critical parts of the orbiter was slim because of the flow of the air stream around the ship during its eight-and-a-half-minute climb to orbit. Even if large chunks of a ramp did hit, the material was lightweight and low-density—like Styrofoam. The conventional wisdom that had evolved from the very first shuttle flight and now pervaded the program was that foam was little more then a maintenance and paperwork hassle—not an issue of life and death.

"We had this mindset that this was a nuisance, maintenance," veteran shuttle manager Hale said months later. "It wasn't a safety-of-flight issue."

Three days after that October 2002 launch, Page and the Kennedy film review team learned just how close Atlantis had come to disaster. As usual, a pair of recovery ships fished the spent boosters out of the Atlantic after their 27-mile plunge from the sky. A quick inspection of the booster struck during liftoff confirmed the foam had hit the attachment ring. The impact had left a crater 4 inches wide and 3 inches deep in insulation covering the ring's forward face. Far more sobering, however, was the discovery that the strike had missed the critical electronics box by less than six inches.

Where the foam had come from remained a mystery until Oct. 18, when Atlantis landed back at Kennedy after an 11-day flight to add a support truss to the International Space Station. Routine photos of the spent tank's jettison snapped by astronauts on Atlantis' flight deck revealed that a corner of the left bipod ramp measuring approximately 4 by 5 by 12 inches was missing. It was only the fourth known example of ramp-foam loss, and the first case in 10 years and 66 launches. After seeing the astronauts' photos of Atlantis' partially missing ramp, managers in the external tank project at Marshall began to talk about possible changes.

One of those managers was Neil Otte, an earnest 44-year-old Midwestern farm boy who was the tank's chief engineer. Otte grew up in the small town of Kahoka, Missouri. After high school, he earned a degree in diesel mechanics from a technical school, then went back to Kahoka and farmed while working on tractors and combines for a John Deere dealer. It took Otte seven years to get bored. He quit his job, sold the farm, and enrolled at Iowa State University, where he earned a degree in mechanical engineering. As graduation neared, Otte sent out résumés to prospective employers, including, on a lark, several NASA field centers. He had little hope of landing a job with the storied space agency, thinking it was beyond the reach of Missouri farm boys like him. To Otte's amazement, an offer came in from Marshall. It was the lowest salary of all his choices, but he took it anyway. He arrived in Huntsville in 1987 as NASA was preparing to return to flight after the Challenger accident. Within a year, he was analyzing stress on the external tank. He rose through the ranks, at one point gaining recognition as the lead engineer for a crucial tank redesign that made it 7,500 pounds lighter and allowed shuttles to carry more cargo to the International Space Station. He was promoted to the tank project office as the lead of the engineering team in 1999, then to chief engineer in 2001. He was named deputy manager—the number two job in the tank project—two weeks after Atlantis' landing.

One option discussed by Otte and other project engineers was removal of a silicon-based coating under the bipod area used to cover the skin of the metal tank before the foam was applied. Known as "super lightweight ablator," or by the acronym SLA (pronounced "slaw") for short, the material looked like chocolate syrup when applied and hardened to the consistency of cork. SLA had been added to certain tank areas for protection against high heating in the early days, but later studies had shown the material wasn't needed in many locations. In fact, SLA already had been removed from several propellant lines and other parts of the tank.

The big worry in the bipod ramp remained the air pockets. Voids often formed between the SLA and the foam. When the fuel tank was filled for launch, the air in those pockets was chilled so deeply that it became liquefied. Then, as fuel levels dropped and the outside of the tank heated up after liftoff, the liquid expanded and became a gas again, creating pressure and popping off pieces of the surrounding foam. Otte and other tank engineers decided the SLA in the ramp area had to go.

The tank project also reexamined the process used to manufacture the ramps. Under the old procedure, a worker at Michoud used a spray gun to manually apply the ramp foam in a single eight-hour shift. There was concern the procedure was responsible for the voids as layers of the rapidly rising foam folded over each other. No test existed to check for voids. Some engineers proposed spraying the ramp more slowly, over several shifts, with the goal of getting more consistent results.

"Everybody had pretty much agreed we would take the SLA off," Otte said. "There was still a lot of discussion in the community about exactly how do we go about improving the [foaming] process."

Shuttle managers convened a meeting of NASA's Program Requirements Control Board on the morning of Oct. 24 to look at a number of technical issues, including the foam strike on Atlantis' booster. The board was charged with deciding whether there were any critical failures during Atlantis' mission worthy of being officially designated in-flight anomalies. The distinction can be an important one. Anything labeled an "IFA" had to be cleared before the next launch and was sure to be widely scrutinized. Page and the Intercenter Photo Working Group recommended that the foam loss on Atlantis' flight be named an IFA, but tank engineers at Marshall thought the designation was unwarranted, insisting that the incident was a freak event.

Page, however, argued the three previous bipod foam incidents, as well as several other "out-of-family" debris strikes, had all been labeled IFAs. He was determined to make a case for what he said was "clearly a dangerous threat to the vehicle."

"It could have been a catastrophic event," Oliu agreed.

"At that moment, it was probably the largest piece of debris we had seen coming off the vehicle. It struck flight hardware, which to us, anything striking a piece of flight hardware is a big deal whether it is a booster, external tank or orbiter. . . . We wanted that item to have the attention of the program, so we pushed it forward with Bob Page."

Several other IFA candidates were on the agenda that day, including a maneuvering thruster that malfunctioned in orbit and problems at liftoff detonating the explosive bolts that hold the boosters to the launchpad. During his film review of the previous mission, Page made his pitch over a speakerphone to the assembled teleconference. He pinpointed where the foam came from and how it hit the booster. He described a pair of photos showing damage to the booster's attachment ring and made it clear how close the impact was to the critical electronics box.

According to an audio tape of the meeting obtained by the authors, a shuttle manager asked Page if anything like this had been seen in the past. "No," he replied, "from my experience, we have not seen impacts like this." A representative from the booster project agreed.

At the end of his briefing, Page summed up the film team's recommendation: Declare the foam strike an IFA "due to the strike area and how early in the flight it was lost" and make resolving the issue a condition for Endeavour's planned November launch.

After Page's presentation, Otte told the board that tank engineers immediately had begun an investigation after seeing the foam strike.

"We've looked at this last tank and we haven't found any process or any anomalies that would have caused us any concern," Otte explained to the group. "However, this is a somewhat difficult closeout [process]. . . . We're saying that from what we've seen so far we probably had a void in there. . . . At about 33 seconds, we had enough internal pressure in the void to blow this off."

Engineers had checked manufacturing records for Atlantis' tank to see if there were any design changes, problems with materials, or processing issues—anything out of the ordinary. They found none. Otte's conclusion: The ramp loss was a unique occurrence and not a generic problem.

"We will go look at if there is a way we can improve this process and make it not as sensitive, but we're still looking at that," Otte told the meeting. "Whether you want to make it an IFA, we're going to leave that up to you."

Jim Halsell, an astronaut who was in charge of Endeavour's ongoing launch preparations at Kennedy, asked "Do we need to look at anything on the pad right now?"

Otte replied, "Right now, Jim, I guess we would answer that no. There is really nothing we can do to that bipod that would give us any more confidence than what we've got right now."

Some of the discussion focused on the definition of an IFA: a critical event that poses a threat to the crew, the shuttle, or mission success. Lambert Austin, a NASA systems integration manager at Johnson, suggested the incident didn't meet the formal definition, though it should be investigated.

"Some said 'No, because it did no damage that was significant to the mission,' " recalled Ron Dittemore, NASA's shuttle program manager and the person who presided over the meeting. "Others said 'That was because you got lucky that day. What if it fell off at 50 seconds and hit some other piece of structure?' . . . Rather than debate that ad nauseum with the two camps, I said 'Table the discussion. Let's work this offline. Let's work it some more. Let's make sure we're smart with what we're talking about. But in the meantime, here are the actions that need to be done.' "

Dittemore was one of NASA's stars, a gifted executive and engineer who had risen up through the ranks and knew the program's day-to-day operations firsthand. Colleagues regarded him as a bright, driven, highly organized manager who occasionally could be hardheaded, arrogant and dismissive of those who didn't measure up.

"He is pretty intense and obviously a real smart guy," said an investigator, who months later would scrutinize the meeting. "There is a dynamic there where his level and his intelligence could have an intimidating effect. . . . You would not want to go to Dittemore without having your homework done. Some of the same qualities that make me respect him so much may actually be what makes him intimidating to some other people."

Dittemore's fascination with space began in May 1961, when, as a 9-year-old Massachusetts schoolboy, he saw Alan Shepard became the first American to venture across the Final Frontier. Soon, he was playing hooky from school to watch the early Mercury launches. In 1963, his father, an Air Force master sergeant, was reassigned to Fairchild Air Force Base in Spokane, Washington, and the family moved west. By then, Dittemore's dream was to one day be part of a human mission to Mars, but he discovered another, more practical, obsession at Medical Lake High School: sports. His favorite was football, but he was talented enough to reach the state tournament in tennis and wrestling. Dittemore graduated third in his class and was named the school's outstanding athlete as a senior. From there, he earned bachelor's and master's degrees in aeronautical and astronautical engineering from the University of Washington, and then took a job with a Phoenix company developing engines for small business jets.

In 1977, Dittemore decided to pursue his lifelong dream by applying for NASA's first class of shuttle astronauts. His application was rejected, but a sharp-eyed flight director pulled his résumé from the stack and asked if he was interested in coming to Johnson and working in Mission Control. Dittemore thought about it and agreed over the phone, without ever visiting Johnson. He accepted a job as a flight controller monitoring the propulsion systems that the orbiter uses in space. During the next decade, he applied five more times to the astronaut corps, taking scuba lessons and earning a pilot's license to improve his credentials. But as his eyesight grew worse, so did his chances. He abandoned the dream in his mid-30s. If he couldn't fly in space, he'd help manage the missions. He was named a shuttle flight director in 1985 and eventually drew assignments working launch and re-entry, the two most critical, high-pressure phases of flight.

In 1992, he left the flight controller ranks and, after a brief stint working on the space station, joined the shuttle program office. Tommy Holloway, a veteran NASA shuttle manager with roots in the Gemini and Apollo era, was impressed and took Dittemore under his wing. After Holloway was named space shuttle program manager in

1995, Dittemore became his integration manager, responsible for the overall coordination of flights, payloads and operations. As part of the job, he also headed the Mission Management Team that oversaw shuttle operations during flight. He managed the shuttle engineering office from 1997 until 1999, when he succeeded his mentor, Holloway, as shuttle program manager.

Dittemore quickly earned a reputation among colleagues and the press as one of the more safety-conscious program managers in shuttle history. While some of his predecessors occasionally were accused of suffering from "go fever" and rushing to launch, he grounded the program twice for months at a time to deal with suspect shuttle wiring and tiny cracks in fuel lines. He raised eyebrows by postponing the December 1999 launch of Discovery to make sure 20-year-old manufacturing records were in order, although there was no evidence the orbiter had a problem. As NASA moved forward with plans for Endeavour's planned November launch, Dittemore's safety credentials appeared impeccable.

He wasn't convinced the foam strike met the definition of an IFA. Even so, he directed the tank project to go to work on eliminating the problem. The tank office would be required to make a presentation on the issue the following week at the Flight Readiness Review for Endeavour's upcoming November mission. That review was a critical meeting where top shuttle managers discussed whether the next mission was ready to safely launch.

"I wanted this briefed at the FRR [Flight Readiness Review] so the entire agency had a chance to listen to the rationale and if they thought it was weak, then they would raise their hands and tell us that we needed to do more work," Dittemore said later. "If the safety organization didn't agree with this, they had an opportunity to raise their hand and say we needed to do more work. That's what a Flight Readiness Review is supposed to be."

Dittemore looked around the room. The teleconference included representatives from safety, the astronaut office, engineering, operations, integration, space and life sciences, contractors, and the shuttle's project offices. He asked one more time if there were any concerns.

"Does anybody else have anything they would like to ask or have done?" Dittemore inquired. There was silence.

Instead of being declared an IFA, the ramp-foam problem was classified under the heading "Plans/Studies." Otte's group was asked to analyze the problem and make recommendations by Dec. 5, almost four weeks after Endeavour's planned Nov. 10 launch. Page went back to his office frustrated, but satisfied that the issue would at least get another hearing at the following week's Flight Readiness Review.

That afternoon, after the board teleconference, Otte and other managers at Marshall gathered to review whether Endeavour's tank was ready to fly. Project officials as well as supporting engineers from Marshall's various labs attended. Their briefing charts said the tank systems had performed normally during the launch of Atlantis. One chart erroneously stated there had been only two previous instances of bipod ramp foam loss—omitting Challenger's 1983 flight—and noted a review of manufacturing and processing records showed nothing irregular. Another chart said analyses of how air currents transported debris coming off the left ramp "showed that foam from this area would not impact the Orbiter." A final chart titled "Rationale for Flight" concluded "Not safety of flight concern."

Not everyone was initially convinced. At the end of the meeting, managers passed around a flight readiness statement, essentially a sign-off sheet, where officials and engineers from the respective departments certified the tank was ready for launch. When it came time for the Structure and Dynamics Laboratory to sign, its representative, Pete Rodriguez, refused. Rodriguez had questions about foam impacts, including several strikes on Atlantis' left wing, which he wanted answered first: How do you know the foam won't come off again? How much of an impact can the orbiter take? How do we know what the orbiter's limits are? A safety manager, Angelia Walker, also refused to sign until Rodriguez's concerns were addressed.

The following day, Otte met with Rodriguez. Otte gave the engineer a brief history of foam loss, explaining how similar events had been handled in the past. The shuttle program never had made foam shedding a constraint to launch before.

Rodriguez signed the flight readiness statement. A technical assistant, Dan Mullane, signed for the safety office.

• • •

One week later, on Halloween, shuttle managers from around the country gathered at Kennedy to attend the Flight Readiness Review for Endeavour's November launch. The meeting occurs before each mission in a large conference room on the ground floor of Kennedy's Operations and Checkout Building, the same historic facility where astronauts since Apollo have suited up before making their celebrated "walkout" amid cheers and flashbulbs to a van that whisks them to the launchpad. For an FRR, senior officials sit at a long table in the center of a cavernous room, flanked on three sides by dozens of other shuttle managers.

The purpose of the FRR is to make sure everything is ready to launch the next mission safely. It is one of the few meetings that virtually all senior NASA officials in the human spaceflight program attend in person. Before the fateful Challenger launch decision, mission managers had argued via teleconference about whether freezing temperatures at the Cape would cause O-ring seals in the shuttle's boosters to fail. After the disaster, some felt the launch decision might have gone differently if the naysayers had been able to make their case face-to-face. For every launch since 1986, the meetings had been attended by all.

This one was chaired by Bill Readdy, the agency's top manager for manned missions. Several technical issues on the agenda had prompted considerable concern. Fuel-line cracks in the shuttles' main propulsion systems had kept the fleet grounded all summer. Although Atlantis had flown in October, it was Endeavour's first launch since being repaired. Engineers were also still trying to fully understand a potentially catastrophic electrical glitch in the system that fires explosive bolts to free the shuttle from the launchpad at liftoff. No one was losing any sleep about foam.

Readdy opened the meeting that Thursday morning by admonishing the gathering to be vigilant about safety. First, there were several presentations on mission operations planned for Endeavour's flight. Then, it was the External Tank Project's turn to talk about the foam loss and other tank issues. The office was represented by its 65-year-old manager, Jerry Smelser. Like Otte, Smelser was a self-described farm boy who grew up in rural northwest Alabama. After graduating from Auburn University with a mechanical engineering degree in 1959, he went to work for NASA when Marshall opened in July 1960 and the space race was heating up in the final months of the Eisenhower administration. Ten years earlier, German rocket scientists, including the legendary Wernher von Braun, had come to the Army's Redstone Arsenal in Huntsville. When Marshall was created from the Redstone Arsenal and von Braun was named its first director, Smelser was one of hundreds of young engineers who were brought in to help build the Saturn boosters that would put Americans on the moon.

"I think there was a quote one time in the *Wall Street Journal,*" Smelser liked to say in his Alabama drawl, "that the Apollo program was successful because of the leadership of a few Germans and because there were a bunch of farm boys from Alabama, Georgia and Mississippi who did not know it could not be done. I am one of those boys."

Smelser spent the next 15 years in the center's Science and Engineering Directorate, moving over to the fledgling shuttle program in 1975. After the Challenger accident, he was named deputy manager of the tank office and later took over the space shuttle main engine project. Following widespread management changes at Marshall in 2000, Smelser agreed to come back as the tank project's manager until his upcoming retirement.

Like the engineers who worked for him, Smelser wasn't impressed with the foam strike on Atlantis three weeks earlier. The same launch photo that had elicited gasps from film reviewers at Kennedy had little effect on him.

"My reaction was that it was similar to a Styrofoam lid of a cooler hitting a windshield of a car or truck," Smelser explained months later. "It was distracting, but not dangerous. We looked at the density of the foam, and it was very light. A piece of foam this size does not weigh much, although I don't have the background or the engineering expertise or all the tools to predict what is going to happen with the foam once it comes off. But from a practical engineer's standpoint, it did not appear to me that anything that light could do damage to an orbiter. That was a practical farm boy engineer's judgment, but that also was substantiated by the people who did the analysis."

Smelser's presentation lasted seven minutes. A videotape of the meeting obtained by the authors shows Smelser standing at a podium, using a pointer to go over slides projected on a giant screen behind him. The first slide showed a photo of Atlantis' tank with the partially missing ramp. Accompanying text declared foam loss "over the life of the Shuttle Program had never been a 'Safety of Flight' issue."

"There are two other known instances where we have lost similar amounts of foam from similar areas," Smelser told the meeting. "Evidence is that was not a substantial contributor to the underside damage to the orbiters."

Smelser moved on to a second slide with pictures of the bipod spindle, before and after being covered with foam. Printed at the bottom in bold italics was the following statement: *"The ET [external tank] is safe to fly with no new concerns (and no added risk)."*

"The rationale for flight is that the tank we're about to fly is equivalent to the tank that we just flew and equivalent to the tanks that we've been flying," Smelser told the assembly. "We have, in fact, a history in this configuration of never causing safety of flight damage on the underside of the orbiter. There are no design changes, process changes or handling of the tank changes that were different about the last tank or that are different about this tank. On the tank that we just flew and the tank that we are about to fly, this operation was performed by experienced practitioners who are very knowledgeable about the application of foam. It is very much an artisanship, craftsmanship operation. . . . The artisanship involved, although the people are well-trained and the process is well-validated, it certainly is dependent on skill—that we have skilled people. The hypothesis—and the very strong belief—about why it [the ramp] came off is there probably was an air bubble somewhere in the foam as it was created."

Smelser explained a bit more about the mechanics of foam shedding, then wrapped up his pitch on the issue:

"Nothing about this tank we are about to fly makes it less probable or more probable of having this happen than the previous one. We've had three out of 112 that we know about. Therefore, we are, in fact, as far as this phenomenon is concerned, we feel it's safe, from a safety of flight standpoint. We can't guarantee that the foam won't come off."

Dittemore recalled later the reaction of shuttle managers wasn't much different from the response in the control board meeting a week earlier. He summed up the general opinion: "I don't think we were impressed with the presentation as being a thorough discussion of the problem."

After Smelser finished, Bryan O'Connor, a former shuttle commander who headed the Safety and Mission Assurance Office at NASA Headquarters in Washington, questioned him on several points. O'Connor asked if the foam size was too small to cause any damage. Smelser conceded "it was a big hunk of foam" but there was no evidence from looking at the orbiter "that this does, in fact, get over and become a safety of flight issue." O'Connor pressed Smelser on whether he was relying too much on past experience. Smelser replied, "There is no evidence that we've ever done safety-of-flight kind of damage. We've done certainly maintenance [damage]."

O'Connor wanted clarification on whether the foam could be dismissed as a safety of flight issue. Recalling *Integrated Hazard Report 37*, O'Connor asked whether it was, in fact, a safety issue but one that had been accepted regardless as a potentially catastrophic risk. Smelser erroneously argued that the report applied only to ice debris. O'Connor asked for the hazard analysis and the discussion was tabled until later in the meeting.

"I didn't want to get into a semantics issue with him," O'Connor said later. "But I thought it was sort of an understated thing to just say that, because to me, this whole topic of stuff coming off tanks and solid rocket boosters and hitting orbiters was an ongoing catastrophic hazard in the program that hadn't been closed out. We haven't made that hazard go away."

A shuttle safety manager brought O'Connor a copy of the hazard report, and later, NASA's safety chief returned to the topic. Despite his desire to avoid a debate on semantics, the discussion boiled down to just that. Smelser conceded that he should have used the phrase "accepted risk to fly" instead of "non-safety of flight issue." But Smelser insisted that his fundamental point was still correct.

"Our review of the integrated hazard report and the ET hazard report seems to be consistent with what I said, with the exception of the fact that I used the wrong words," Smelser said.

With that, the debate ended. Readdy suggested that an action be given to update the integrated hazard report because it hadn't been revised recently and didn't address the particular hazard of ramp foam coming off. Dittemore made a counterproposal that the shuttle program's Safety System Review Panel look over the report to see if it needed technical changes or beefing up.

"I'm not sure it needs updating," Dittemore told the group, "but it sure appears that we need to review it again to make sure it's accurate in its language, given what we've discussed today."

The meeting moved on without taking any formal action. Atlantis' debris strike would pose no roadblock for Endeavour's mission—just as foam never had held up a flight in the past. However, NASA would make sure its paperwork was in order. The hazard report would be looked over to be certain its wording more precisely reflected the nature of the foam threat.

Endeavour's launch date was set for Nov. 11. The mission was delayed during the final stages of the countdown, however, by a leaky oxygen hose in the shuttle's midsection. While attempting to fix the problem, workers installing an access platform in Endeavour's cargo bay banged the shuttle's robot arm, ripping part of its protective thermal insulation. Engineers meticulously studied the arm for damage, while Dittemore held up the flight for more than a week to make certain the oxygen leak was not a generic fatigue issue affecting hoses throughout the shuttle fleet.

Both issues finally were cleared, and Endeavour thundered off at 7:50 p.m. on Nov. 23. Because this was a night launch, it was impossible to tell whether the foam ramps on Endeavour's tank remained intact during the climb to orbit. But just as Smelser and the tank project had predicted, the ship suffered no serious foam damage. Endeavour landed safely back at Kennedy on Dec. 7 after successfully ferrying a new crew to the space station. Inspections after landing showed foam dings to tiles on the shuttle's belly all were "in family." It appeared the first major loss of bipod ramp foam in 66 missions had been a fluke.

Meanwhile, Smelser's team at Marshall had fallen behind schedule in reporting back on possible fixes for the ramp problem. The group asked for an extension beyond its Dec. 5 deadline. Dittemore agreed. Their report eventually was shifted to the first week of February, the week after Columbia was scheduled to land following the next shuttle flight.

"They came in and said 'Look Ron, we don't have anything to really tell you on Dec. 5,' " Dittemore remembered. " 'Why don't you give us another six weeks or something like that?' I said 'Okay, I want you to be right rather than wrong. I want to have a meeting of substance rather than you just giving me a bunch of viewgraphs.' So I said I would give them another six weeks. Did I think it was an FRR question? No."

Senior shuttle managers convened at the Cape again on Jan. 9 for Columbia's Flight Readiness Review. To most officials, Atlantis' debris strike two flights earlier was another forgotten chapter in the constantly growing annals of foam loss. The issue never came up.

Dittemore, however, would be haunted in the coming weeks and months by Endeavour's readiness review the previous October.

"I look back on it now and say that was one of the things we really missed as an agency," Dittemore said. "We had a presentation that was not thoroughly discussing the risk. And four center directors, program management, contractors, and NASA—the associate administrator for spaceflight and the associate administrator for safety for the agency—said it was okay."

"Looking back on it, that was a critical juncture where we as an agency—at the highest levels of our headquarters management and our program management and our contractor management and our team sitting in that room—had a problem that was not thoroughly discussed. And we pressed forward."

FOUR

LAUNCH

Experts have reviewed the high speed photography and there is no concern for RCC or tile damage. We have seen this same phenomenon on several other flights and there is absolutely no concern for entry.

—*Flight director Steve Stich,*
in an e-mail to Columbia's crew about
the launch-day foam strike

Columbia's 5-mile-per-second voyage began with a glacial 9-inch-per-second trip from Kennedy's cavernous Vehicle Assembly Building atop a squat Apollo-era crawler-transporter. It was Dec. 9, 2002, and it took six hours for the massive crawler to carry Columbia and its mobile launch platform—an 11-million-pound load—to the launchpad just over three miles away.

Husband and his crewmates flew to Kennedy nine days later on December 18 to review emergency escape procedures and to participate in a dress-rehearsal countdown, a major milestone on the road to launch. Ramon's presence on the crew generated intense media interest, much of it focusing on NASA's security precautions. The agency already had tightened up security in the wake of the Sept. 11, 2001, terrorist attacks on the World Trade Center and the Pentagon, and officials said no major changes had been ordered for Ramon's launch. But as Columbia's launch date approached, a heavier than usual police presence was noted, and at the gates leading into Kennedy, guards carrying automatic weapons seemed to give everyone's security badge closer scrutiny than usual.

"Based on the world climate today, it would be foolish to try to make you believe that the presence of an Israeli astronaut on this flight would not make it a higher profile flight," David Saleeba, a former Secret Service agent and chief of NASA security, told reporters. "It's unfortunate that is true; it shouldn't be. He's first and foremost an astronaut and a scientist. And on a secondary note, he's Israeli. We fly people from all over the world on many of our flights."

For his part, Ramon said he had no concerns. "I think NASA security is doing everything needed since Sept. 11. I feel great safety," he said. "I think everybody feels safe."

On Friday, Dec. 20, the astronauts donned their pressure suits and climbed into Columbia for their practice countdown. The procedure was identical to what the crew and the launch team would experience during the actual countdown. The astronauts flew back to Houston that afternoon for the holidays. The next day, the Clark family took another flight that almost ended in tragedy in the skies above Texas.

"It was a Bonanza, single-engine, high-performance airplane," Jon Clark recalled. "We hit turbulence like I've never seen. We were getting beat all over the sky. [Laurel] actually got sick, which is pretty rare because she was not prone to motion sickness and she'd flown in a lot of military aircraft and been at sea. So it was pretty bad."

He descended to 500 feet to inspect a rural runway before attempting an emergency landing: "All of the sudden we hit this, like a downdraft, and that was it, just right toward the ground. . . . So I'm just trying to basically control the crash. She's in the back and I'm like, I don't know what's going on here but basically, you know, this is not good."

The plane crashed just off the runway, slamming down hard enough to destroy the Bonanza. Miraculously, no one was injured. "But it was pretty upsetting," Jon said. "It was really upsetting to our son."

Columbia's formal Flight Readiness Review was held Jan. 9, 2003, at Kennedy. Foam shedding was never mentioned, but lingering questions about a problem with potentially cracked bearings in the fuel lines remained unresolved. The review concluded with tentative clearance to launch Columbia on Jan. 16, between 10 a.m. and 2 p.m., assuming engineers could complete final tests showing the bearings were not a safety-of-flight issue. In keeping with post-9/11 security procedures, the exact launch time would remain secret until 24 hours before liftoff.

The astronauts flew back to Kennedy the evening of Sunday, Jan. 12, arriving a few hours before the start of the countdown at 11 p.m. The countdown was longer than usual because of time needed to load extra liquid oxygen and hydrogen on board to power the ship's three electricity-producing fuel cells. The fuel cells combined hydrogen and oxygen to produce electricity and, as a by-product, fresh water. To carry out the planned 16-day mission, Columbia was carrying an extra set of oxygen and hydrogen tanks at the back of the cargo bay.

The astronauts spent their time reviewing the flight plan, taking spins in jet trainers, attending a series of readiness briefings, and monitoring the weather. Their training was over. This was a time to collect one's thoughts and to spend a final few moments with immediate family members before strapping in for the launch. The astronauts already were in medical quarantine, and the number of people allowed to have personal contact with them was strictly limited.

The families flew to Kennedy on Monday, and checked into motels. The Ramons, along with scores of invited Israeli guests and VIPs, including Daniel Ayalon, Israel's ambassador to the United States, stayed at a Hilton Hotel in Cocoa Beach 20 miles south of the space center. Barricades and a police command post were set up in the parking lot, and no one got in without proper identification.

"We went down there on a Monday, and every day after that, we had time with the crew," Jon Clark said. "The kids couldn't go because they were kind of excluded by the quarantine. But we had a lot of time with them, spent some precious days with them. It was hard on [Iain] because he knew she was there and that I was seeing her. I was trying to keep it real low key."

The Sunday before the astronauts had left Houston, Clark had broken the crew's quarantine at Johnson to give Iain a final moment with his mother.

"I snuck him in the back and he hugged her even though it was in the seven-day period," Jon said. "I just said screw it, you know?"

Columbia's countdown ticked along without any major glitches. Ideal weather was expected on launch day. A final readiness review was held the day before the launch, and Dittemore formally put the bearing issue to rest, agreeing with the engineering community that Columbia could be safely launched. Tests had shown that even if any of the 16 bearings in the shuttle's propulsion system were cracked, the odds of a large chunk of metal breaking off and damaging an engine were minimal.

That afternoon, a protective gantry was rolled away and Columbia stood exposed to view, ready for the terminal phase of the countdown. The sun soon set, and powerful xenon spotlights fired up, bathing the shuttle in shafts of light visible for miles around. Just before 11 p.m., engineers and technicians began evacuating the launchpad. The Mission Management Team, chaired by Ham, met a half-hour later to assess the weather and the progress of the countdown.

Last-minute concern about the structural strength of the massive attachment rings that help hold the bottom of a shuttle's boosters to the external fuel tank had surfaced earlier in the day at Johnson. The concern, which was not related to Atlantis' debris strike two missions earlier, was based on test data indicating the material used in the aft attachment rings was not quite as strong as required by NASA. Should a ring fail in flight, the base of a booster could pull free of the tank in a repeat of the Challenger disaster.

At Columbia's tanking meeting, engineers presented a last-minute analysis to Ham that showed the rings were safe for flight. The management team signed off on the issue, and the launch team was given permission to begin loading oxygen and hydrogen fuel into Columbia's external tank.

The three-hour process began a few minutes after 3 a.m. on Thursday the 16th, about an hour and 10 minutes behind schedule because of work to fix a handful of minor problems. Operating by remote control, engineers loaded the upper section of the giant tank with 143,000 gallons of liquid oxygen, while the larger, lower section was filled with 385,000 gallons of volatile liquid hydrogen. Unlike an automobile, which burns gasoline with oxygen from the atmosphere to provide energy, Columbia had to carry its oxygen with it.

The propellants entered the shuttle through panels on each side of its aft compartment, flowing through each engine to chill the hardware and its plumbing before flowing out through 17-inch-wide pipes in the belly of the shuttle. From there, the hydrogen flowed directly into the bottom of the external tank while the oxygen was pumped up a long pipe to the tank's upper section.

By 5:50 a.m., fueling was complete. Members of the ice, debris, and final inspection team that Bob Page and Armando Oliu belonged to drove to the launchpad to search for potential hazards. They were primarily interested in looking for signs of dangerous ice buildups on the external tank that could break off during launch and strike the shuttle's tiles. They also made a last-minute check to make sure no potentially dangerous debris had been inadvertently left on the pad that could pose a threat to the orbiter. The team reported a slight buildup of frost on the tank's left bipod ramp, but that wasn't unusual and no other problems were noted.

The astronauts, meanwhile, were suiting up for flight in their quarters at the Operations and Checkout Building. McCool, Anderson and Brown, who would work the overnight or "blue shift" once in orbit, had been up all night, monitoring the fueling process. Husband, Chawla, Clark and Ramon, the day or "red shift," were awakened just before sunrise. All seven astronauts attended a final weather briefing and checked in with flight controllers in Mission Control at Johnson for a last-minute update.

A team of technicians helped them don their 70-pound pressure suits, making sure the internal headphones, microphones, and air supply systems were working properly. It was time to head for the launchpad. As they walked down a hallway to the elevator that would carry them to the ground floor, Husband stopped and the crew huddled together.

"They all gathered in a circle and bowed over and put their arms around each other and they said a prayer together, all of them," said astronaut Bob Cabana. "I mean we have Christians, Jews, Hindus. . . . This crew was bound as one at that point."

A throng of Kennedy workers, reporters, and news photographers were gathered outside crew quarters to give the astronauts a final send-off. Husband and McCool walked out first, smiling, waving and reaching up to pat an STS-107 mission patch glued to the door frame. Chawla, Clark, Ramon, Brown and Anderson were right behind, in equally good spirits. They climbed into NASA's waiting "Astrovan," a modified motor home used to carry astronauts to and from the launchpad, and departed shortly before 7:30 a.m.

Overhead, unseen by the astronauts or anyone else in the crowd waving good-bye, fighter jets were patrolling an FAA-mandated no-fly zone extending 30 miles in all directions. A powerful military radar system and reconnaissance planes were scanning the skies, and the Coast Guard was patrolling offshore. Local airports were shut down, and mobile missile batteries reportedly were in place at the nearby Cape Canaveral Air Force Station, ready to shoot down hostile intruders. NASA and the Air Force were taking no chances.

The Astrovan stopped briefly at the Launch Control Center to drop off Cabana and Kent Rominger. Cabana was heading for the firing room. Rominger was heading for the nearby shuttle runway to begin landing runs in a Shuttle Training Aircraft, the same jet he would fly 16 days later during Columbia's re-entry. If an engine failed early in flight, Husband and McCool would have to attempt an emergency return to the launch site. Rominger's job on launch day was to assess the landing conditions.

"I shook everybody's hand and gave everybody a hug, Rick last, and said goodbye to them," Cabana said. "I told him to be safe and have a great flight."

The Astrovan headed on toward the pad three miles east, preceded by a low-flying NASA helicopter. Armed security guards were sitting in its open doors, automatic weapons at the ready. Reporters across the road waved as the Astrovan pulled away, a launch-day ritual of sorts for many.

Ten minutes later, the Astrovan pulled up to a strangely deserted launchpad. Normally bustling with activity, the pad was evacuated on launch day because of the danger inherent in the external tanks explosive hydrogen. Only a half-dozen workers were there to greet the astronauts, a small team responsible for strapping them in and closing the shuttle's main hatch.

Husband and his crewmates rode an elevator to the 195-foot level of the main tower. They paused briefly at the top, taking in the view and listening to the creaks and groans of the frigid tank. Husband didn't tarry. He walked across an enclosed walkway that extended to the side of Columbia's crew module and entered the "white room," a cramped enclosure that protected boarding astronauts from the weather.

He was first aboard, crawling through Columbia's side hatch on hands and knees, followed by Ramon and McCool a few minutes later. Brown was next in line. For launch, he would be seated on the flight deck and would trade places with Clark for landing. Before boarding, he made a quick phone call to Ann Micklos.

"He called me from the 195-foot level and left me a phone message, which I did not get until that evening after launch," Micklos said. "[He's] just thanking me for all that we had done to prepare the vehicle and he can't believe that he is finally up and it is his turn to get strapped in and just thanks again for all the hard work and looking forward to seeing me when he got back."

Brown was followed by Clark and, finally, Chawla. The closeout crew shut and locked Columbia's hatch at 9:17 a.m., almost exactly the shuttle's projected landing time 16 days later. After two-and-a-half years of training and frustrating delays, Husband and his crewmates were finally braced for launch on the 113th shuttle mission. The endless simulations and practice runs were over. The crew's big moment was finally at hand.

The countdown ticked smoothly toward zero. At the end of a final "hold" at the T-minus-nine-minute mark, launch director Mike Leinbach called Husband with an update.

"OK Rick, if there was ever a time to use the phrase 'all good things come to people who wait,' this is the one time," Leinbach told Husband from a console in the Launch Control Center. "For you and your crew, best of luck on this mission and from the many, many people who put this mission together, good luck and Godspeed."

"We appreciate it, Mike," Husband replied, sounding relaxed. "The Lord has blessed us with a beautiful day and we're going to have a great mission. We appreciate all the great, hard work everybody's put into this and we're ready to go."

At T-minus 31 seconds, Columbia's four flight computers took over the final moments of the countdown. Only a handful of missions had been derailed this close to launch and for Husband and his crewmates, the moment of truth was at hand. The external fuel tank had been pressurized, the shuttle's hydraulic system was up and running, the fuel cells were generating power and the boosters and main engines were primed for ignition. There were no problems. It was time to go.

". . . 10, 9, 8, 7, we have a go for main engine start . . ."

Ten stories below the astronauts, the shuttle's three main engines ignited with a muffled roar, firing up at 120-millisecond intervals. Within seconds, they were producing 37 million horsepower, the energy output of 23 Hoover Dams. In the cockpit, the astronauts felt the spacecraft shake and rattle to life as the engines throttled up to full power.

". . . 5, 4, 3 . . ."

The thrust from the slightly canted engine nozzles pushed Columbia's nose toward the Atlantic Ocean, and the tip of the ship's external tank moved about 3 feet before the vehicle rebounded back toward vertical. The hot, transparent exhaust from the engines instantly vaporized torrents of water flooding the surface of the launch platform, sending billowing clouds of steam into the blue January sky.

Three miles away, standing on the roof of the Launch Control Center, the families of Columbia's astronauts watched with nervous anticipation. It would take several seconds for the sound to arrive. Until then, they hung onto every word from NASA public affairs commentator Bruce Buckingham, who was calling out the final moments of the countdown from Firing Room 3 two floors below.

The shuttle's computer system, meanwhile, was checking the operation of each main engine 50 times per second, ready to automatically order shutdown, stopping the countdown if even the smallest valve failed to work properly. McCool was a final human link in the chain, staring intently at a computer screen directly in front of him that verified all three engines had spun up to 100 percent thrust.

". . . 2, 1 . . ."

With the engines generating more than a million pounds of push, Columbia's four flight computers finally issued the only irreversible commands of the countdown, the ones that would actually launch the spacecraft. The signals raced through the shuttle's wiring to small rocket motors buried in the forward sections of Columbia's two solid-fuel boosters. Tongues of flame shot down into the hollow core of each 12-foot-wide booster, igniting their 1.1-million-pound loads of rubbery propellant. In the blink of an eye, pressure inside each 149-foot-tall solid rocket booster, or SRB, jumped from sea level—14.7 pounds per square inch—to more than 900 psi. Twin plumes of 5,000-degree exhaust shot out the bottom of each booster, providing an additional 5 million pounds of thrust.

". . . we have booster ignition . . ."

At that same instant, explosive charges were triggered inside the four 7-inch-wide bolts at the base of each booster that, until this moment, had held the boosters to the launch platform. In a split second, the charges detonated and the bolts blew apart. The seven astronauts were rocked and shaken in their seats as the boosters unleashed their fury, and Columbia began its eight-and-a-half-minute climb to orbit.

". . . and liftoff of space shuttle Columbia!" exclaimed Buckingham, his words going out over NASA's television network.

The astronauts felt a near-instant rush of vertical acceleration, some of them perhaps surprised by the violence of the ride. Husband saw the launchpad tower fall away outside his left window. The computers ordered a slight increase in the engines' throttle settings, and in less than 10 seconds, the spacecraft had accelerated straight up to more than 100 mph, quicker than the commander's vintage Camaro could manage on level ground.

The sound of the engines finally reached the Launch Control Center, overwhelmed seven seconds later by the crackling roar of the boosters. The families standing on the roof felt it shake under their feet as if a small earthquake had struck. They yelled in excitement, hugging one another and clapping their hands. But they could not even hear their own cheers.

For the first 10 seconds, the computers kept the booster and main engine nozzles locked in place and the shuttle climbed straight away from the pad. As soon as the ship cleared the launchpad gantry, however, hydraulic pistons were ordered to push the nozzles at the base of each booster in opposite directions. The spacecraft began rolling around its vertical axis and leaning over, arcing to the northeast and lining up on a trajectory tilted 39 degrees to Earth's equator.

"Houston, Columbia, roll program," Husband radioed.

"Roger roll, Columbia," replied astronaut Charlie Hobaugh in Mission Control.

The effect of the seven-second roll maneuver was to put Columbia underneath the fuel tank with the astronauts inside sitting upside-down as the spacecraft crossed over the Atlantic Ocean.

McCool and Husband, with Chawla looking over their shoulders, kept their eyes on the dashboard, watching their computer screens for signs of trouble. The shuttle's flight computers were doing the actual flying, displaying the ship's velocity, its orientation, and details about its trajectory and engine performance on the forward cockpit displays. The crew was strictly along for the ride. Husband could take over manual control in an emergency, but that was considered a desperate last resort.

The astronauts did not bother worrying about a booster failure. Once ignited, the SRBs burned until their fuel was exhausted and major malfunctions were not considered survivable. The liquid-fueled main engines were another matter. The computers could shut an engine down, if necessary, and the crew would have to be quick to respond.

An engine failure early in flight would force them to attempt an emergency return to Kennedy, a hair-raising abort scenario requiring the shuttle to flip around and briefly fly backward at five times the speed of sound. Failures later in flight could force the crew to land in Spain or to limp into a lower-than-planned orbit, something that had happened once before in 1985. Husband and McCool had spent endless hours practicing abort procedures they hoped would never be needed. So far, the engines were operating flawlessly.

Right on time, a half minute after liftoff, the computers began throttling the main engines down to 74 percent power to ease the aerodynamic stress on the vehicle as it rocketed through the sound barrier. The engines started throttling back up 12 seconds later, prompting a familiar call from Mission Control: "Columbia, Houston. Go at throttle up." The call told Husband that as of this moment, flight controllers saw no problems with any of Columbia's major systems.

The shuttle was getting lighter by the second, consuming fuel at enormous rates and steadily accelerating. In the first 80 seconds, Columbia burned up 2 million pounds of solid and liquid propellant reaching an altitude of 12 miles and a velocity of 1,500 mph.

Back on the roof of the Launch Control Center, the thunder was fading. Columbia was arcing away to the east, leaving behind a long, churning trail of exhaust. Challenger had been destroyed by a booster failure 17 years earlier, and so everyone breathed a sigh of relief when Columbia's boosters, their fuel exhausted, fell away at an altitude of 27 miles.

On the flight deck, the astronauts saw a sudden burst of orange fire and smoke wash over the cockpit windows as small rocket motors ignited to push the spent boosters away. The shaking suddenly stopped, and the ride became glassy smooth. Accelerating through 4,000 mph—twice as fast as a rifle bullet—on the power of its three main engines, the craft and its white-hot exhaust looked like a brilliant star in the daytime sky.

At the Launch Control Center, Noa Ramon, the Israeli flier's young daughter, was crying. "I lost my daddy," Jon Clark heard her say. Her mother, Rona, reassured her that everything was all right and that her father would be back in 16 days. "No, I've lost my daddy," Noa insisted. Jon and his son Iain were standing off to one side. Iain was crying uncontrollably.

But there were no signs of anything amiss. Five minutes into the climb to orbit, Columbia was high enough (66 miles) and going fast enough (7,500 mph) to make it into orbit even if a main engine failed. The shuttle was flying almost horizontally now, trading altitude for speed, and veteran shuttle watchers began relaxing a bit. A few seconds later, the ship's flight computers carried out another roll program, this one to put the shuttle on top of the external tank to permit better communications through a NASA tracking satellite.

The acceleration mounted. Soon, the astronauts were being pushed back in their seats as if they weighed twice, then three times their normal weight. At seven minutes and 21 seconds into flight, Columbia's computers began throttling down the main engines to prevent the loads on the crew and the shuttle from exceeding three times the force of gravity.

One minute later, the flight computers ordered valves in the shuttle's fuel lines to close, and the engines shut down. The astronauts were thrown forward against their straps and small items began to float about the cabin. Columbia was in space.

Husband reached over and shook hands with McCool, now an experienced shuttle pilot. Columbia's velocity was 17,600 mph in a preliminary orbit with a high point of 194 miles and a low point of 55 miles. In 44 minutes, on the other side of the planet, Husband and McCool would fire the shuttle's maneuvering rockets to circularize the orbit at an average altitude of 184 miles.

It was all standard procedure. But first, they had to get rid of their now empty external fuel tank, the only part of the shuttle system that is not reused.

Seconds after main engine cutoff, explosive charges detonated, severing the tank's connections with Columbia. Small downward-pointing steering jets then fired to push the shuttle up and away from the tank. Motorized doors swung shut over the massive connections in the belly of the orbiter where the tank's fuel lines had entered the engine compartment. The shuttle maneuvered to give the astronauts a view of the tank through the cockpit's two overhead windows.

Brown, sitting to Chawla's right, quickly unstrapped and pulled out a video camera. Anderson floated up from the lower deck carrying a still camera with a telephoto lens. Brown's first job in orbit was to photograph the external tank before it fell back into the atmosphere. Engineers would study the film after landing to look for signs of foam shedding and to make sure their analysis of the bipod incident two flights ago in October had been correct. They had reason for optimism. Film shot by the crew of shuttle Endeavour in November didn't show any evidence of major foam loss, and in another few missions, the bipod changes being studied in the wake of the October flight would likely be in place.

But by the time Brown began videotaping Columbia's tank, the bipod area had rotated out of view. If he saw anything unusual before then, he never reported it. As far as the crew was concerned, launch had been flawless. Columbia was in orbit, and it was time to get to work.

From a management and crew scheduling point of view, Columbia's mission was one of the most complex shuttle flights attempted in recent years, a mix of more than 80 government, university, and commercial experiments from researchers in the United States, Canada, Japan, the European Space Agency, and Israel. NASA's public affairs staff churned out dozens of fact sheets, color brochures, and feature stories touting the scientific value of the research.

But it was a tough sell. The experiments seemed second-tier to many outside observers. It was not that any one experiment represented demonstrably bad science. But given the half-billion-dollar cost of a shuttle flight, critics argued the price tag far outweighed the potential benefits.

Even some in the shuttle program privately questioned the value of Columbia's research. The mission lacked the sort of central theme that defined earlier shuttle research flights. Put on the manifest at the insistence of Congress, Columbia was loaded with a wide range of unrelated experiments in multiple disciplines, defying easy explanation. Shuttle research missions typically had featured experiments with more tightly focused objectives, such as astrophysics or the effects of weightlessness on human physiology, materials processing or technology development.

It wasn't just outsiders who had problems getting a grip on Columbia's research payload. Mission managers couldn't even agree among themselves on how many individual investigations were on board. Reporters settled on a vague "more than 80." Mission managers did, however, insist publicly that Columbia's flight was scientifically significant and a potential boon to humanity. Left unsaid was why the flight had the lowest priority on the shuttle launch schedule and why it was repeatedly shuffled to the back of the queue. Columbia's frequent delays made it painfully clear where the mission stood in terms of NASA's priorities.

Such considerations were a moot point to Husband and his crewmates, who clearly believed in what they were doing and were willing to put their lives on the line to ensure success.

"A lot of hard work has gone into preparing these experiments. People have taken great pains to make sure that this science is valuable and worthwhile," Anderson said before launch. "We're just excited to be able to take them up there and, hopefully, bring back some of the best science we've had in years."

But their enthusiasm could not offset the widely held belief among shuttle critics and other outsiders that Columbia's mission was rooted more in politics and Spacehab's lobbying prowess than in pure science.

Most of Columbia's experiments were housed inside a module built by Spacehab as a commercial venture. Twice the size of Spacehab's normal space station cargo modules, the so-called research double module was 20 feet long, 14 feet wide, and 11 feet high. Mounted in the central portion of the shuttle's cargo bay, it was connected to the crew cabin by a pressurized tunnel. The module was roomy enough to accommodate three or four astronauts at a time and the crew enjoyed performing zero-gravity gymnastics from time to time, doing somersaults or spinning about in the center of the laboratory as time permitted.

The module could carry up to 9,000 pounds of cargo and experiment hardware but for its debut flight aboard Columbia, it was loaded with about 7,500 pounds of equipment and supplies. NASA booked just over 80 percent of the available space for agency-sponsored experiments. The rest was sold by Spacehab to government agencies and commercial users around the world.

For Columbia's flight, the module was loaded with nine commercial payloads supporting 21 investigations into bone loss, protein crystal growth, the effects of weightlessness on human physiology and technology demonstrations in space navigation, satellite communications and thermal control systems. Another four payloads provided by the European Space Agency supported 14 other investigations. And NASA was responsible for a space station technology development experiment, along with 18 payloads sponsored by the office of Biological and Physical Research for 23 investigations.

As with previous shuttle research missions, virtually all of the experiments were designed to shed light on processes that often are masked or otherwise difficult to observe because of the effects of gravity on Earth. By conducting the experiments in orbit, gravity could be taken out of the equation, perhaps giving scientists insights into fundamental processes in biology and the physical sciences. Whether those insights could be translated into Earthly benefits wouldn't be known until the research was carried out.

"In the areas of biology and physiology, we'll be conducting several studies to study the effects of weightlessness on crystal growth, plant growth and also the cardiovascular and musculoskeletal systems," said Kelly Beck, lead flight director for mission STS-107. "The information we gain from these studies will help develop better drugs, with particular interest in the areas of cancer and osteoporosis research.

"We also have studies of weightlessness on the human body, which will lead a better understanding of the effects of spaceflight and help us develop better techniques to combat those effects and increase our durations in space," Beck said. "There are studies of combustion and fire suppression processes that will help us potentially lead to cleaner burning techniques here on the ground as well as improved fire suppression and prevention techniques."

Three technology demonstration experiments were mounted on the roof of the Spacehab module, exposed to the harsh environment of space. Mounted directly behind the Spacehab module was a pallet of experiments known by the acronym FREESTAR. One experiment was devoted to studying the behavior of xenon at low temperatures, another was designed monitor Earth's ozone layer, and a third measured the amount of solar radiation reaching Earth, part of an ongoing study to chart the long-term value of the sun's energy output. The FREESTAR suite also included a package of 10 student experiments in the Space Experiment Module; an experimental space radio; and Ramon's Mediterranean Israeli Dust Experiment, or MEIDEX.

"We also have several experiments to demonstrate new technologies, such as water recycling, thermal control, communications and navigation," Beck said. "So that just kind of gives you a flavor of the wide variety of things we'll be doing on this mission."

Six experiments inside the Spacehab module were provided by students in a half-dozen countries as part of an education program managed by Space Media, a Spacehab subsidiary. Student experiments are secondary objectives on shuttle missions, but they receive a disproportionate share of the publicity because of their more easily understood objectives and because they frequently involve plants and animals. Columbia carried 13 rats, eight garden orb weaver spiders, five silkworms and three cocoons, four Medaka fish eggs, three carpenter bees, 15 harvester ants and an assortment of fish, mostly because students wanted to see how they would behave when weightless.

One student experiment later attracted potshots from critics. Called "Fun with Urine," the idea was to test the feasibility of urine-based paint as a possible way to redecorate future spacecraft, and thus stave off depression, on long-duration voyages.

• • •

It would take Columbia's crew 16 days to complete all of its research. Four hours after reaching orbit, McCool, Anderson, and Brown went to bed, slipping into coffin-like sleep stations on the right side of the lower deck that could be closed off for privacy. While they slept, Husband and the red shift set up Columbia for orbital flight, activating the Spacehab module.

"The vehicle just flew beautifully, we had a great ride," Husband later told flight controllers, "Some people have said Columbia, they have to do a lot of work on it on the ground, but when it gets into space, it just does wonderful. And we think that's because she just loves being in space."

Because many of the experiments involved transmitting data back to Earth through the shuttle's high-speed communications system, the astronauts were not able to beam down as much live television as other crews and only a handful of media interviews were scheduled. The first two took place two days after launch when Husband, Chawla, Clark, and Ramon fielded questions from CBS News and the Cable News Network (CNN).

"Things are going really well," Husband said. "Columbia is in great shape and working absolutely perfectly, the experiments are working very well also. So I'd say we're batting at least a thousand."

Ramon said the flight was off to a "great" start. "It's an opening for great science from our nation and hopefully, for our neighbors in the Middle East," he said. He described the launch as "pretty exciting, yes, with a lot of noise and shaking." Ramon fielded questions about how he planned to observe Jewish customs in space, where shuttle crews experience 16 sunrises and sunsets a day. He said he'd been so busy during the first two days of the mission that he had forgotten to mark his first Sabbath in orbit. Before the launch he had described himself as a secular Jew, but he promised to observe what traditions he could as a representative of Israel. He brought kosher food with him, along with Kiddush cups and several cases called mezuzahs containing small scrolls with inscriptions from the Book of Deuteronomy. As it turned out, however, he didn't have to worry about observing the Sabbath more than once a week. It was decided before launch that he would mark the Sabbath in orbit when it was observed at the launch site in Florida.

Clark commented on her delight that "obviously, everything floats" in space. "The zippers and all the belts that have D-rings that we hold things down with are always floating and hitting each other and jingling. And its makes this beautiful tinkling music in the background all the time," she said. "It just caught me off guard and it was beautiful."

Jon Clark said his wife had the same reaction all rookie fliers have to the novel experience of weightlessness. "There's a kind of a space euphoria that hits these guys and they're just bubbly and excited," Clark said. "She was just ebullient, just in awe of the experience. You got the sense of just incredible satisfaction and joy."

Five days into the mission, on Tuesday, Jan. 21, the day after the Martin Luther King Day holiday, veteran shuttle manager Phil Engelauf briefed reporters on the progress of Columbia's mission.

"The orbiter and the crew are all doing very, very well," he said. "We have no significant anomalies to report." The only issue even worth mentioning, he said, was trouble with an air conditioner in the Spacehab module. Water was leaking out of a humidity collection system, and the crew had been asked to put up with slightly higher temperatures in an effort to reduce condensation. In the weightless environment of space, free-floating water was potentially an electrical hazard. But in this case, it was more of an annoyance than anything significant.

"That's really the only issue we've got to talk about in terms of orbiter or Spacehab systems," Engelauf said, "and it doesn't look like it's going to be an impact to science."

The mission settled into a comfortable routine. News coverage of the flight dropped off, and few outside the space community paid much attention.

Later, rumors started to ripple through the space agency about a debris strike of some sort during launch. Several reporters, including the authors of this book, either called NASA public affairs with questions or were contacted by agency sources alerting them to the incident. Public affairs told the few reporters who called that engineers had assessed the strike and concluded it was not a threat.

The astronauts, however, knew nothing about the impact. One week into the mission, flight director Steve Stich sent Husband an e-mail alerting him to the issue because the management team didn't want the crew to get blindsided by questions in upcoming interviews.

> You guys are doing a fantastic job staying on the timeline and accomplishing great science. Keep up the good work and let us know if there is anything that we can do better. . . .
>
> There is one item that I would like to make you aware of [before the planned interviews]. This item is not even worth mentioning other than wanting to make sure that you are not surprised by it in a question from a reporter.
>
> During ascent at approximately 80 seconds, photo analysis shows that some debris from the area of the [left-side external tank] Bipod Attach Point came loose and subsequently impacted the orbiter left wing . . . creating a shower of smaller particles. The impact appears to be totally on the lower surface and no particles are seen to traverse over the upper surface of the wing. Experts have reviewed the high speed photography and there is no concern for RCC or tile damage. We have seen this same phenomenon on several other flights and there is absolutely no concern for entry.
>
> That is all for now. It's a pleasure working with you every day.

FIVE

A SHOT IN THE DARK

Why is Gene Kranz a hero? Well, Apollo 13. I mean, they had the greatest catastrophe that had ever happened in space, and it was the greatest day ever for Mission Control because they saved the crew and snatched victory from the jaws of defeat. . . . I can't say that. We never even gave ourselves the opportunity.

—*Wayne Hale, NASA launch integration manager and veteran flight director*

A terrible sense of déjà vu came over members of Kennedy's film review team as they saw the first images of Columbia's climb to orbit.

A call had come in from an alarmed Marshall engineer around 10 a.m. that Friday as he and a film team in Huntsville screened images of the previous day's liftoff. Low-resolution television video from an Air Force tracking camera 26 miles south of the launchpad in Cocoa Beach had captured a large debris strike. Coincidentally, the team at Kennedy had just loaded a sharper 35-mm view from the same camera location that had arrived a half-hour earlier from the developer in Miami. The camera, however, had been out of focus. The film was virtually worthless.

Armando Oliu, Bob Page, and other Kennedy engineers quickly cued up film from another high-resolution camera 17 miles away at a Cape Canaveral Air Force Station runway. They too were stunned by what they saw.

A large chunk of foam had broken away from somewhere in the bipod area of Columbia's external fuel tank 81 seconds after launch and smashed into the ship's left wing near the leading edge, producing a spectacular shower of particles. It was unclear from the images whether the debris cloud contained foam, ice, or bits of Columbia's heat tiles. Exclamations of "Holy shit!" and "Oh, my God!" filled the film lab. This time, however, the impact wasn't on a booster. It was on the orbiter itself.

"There were some team members who had been here since STS-1 [the first shuttle launch in 1981]," Oliu said. "By far, it was the largest piece of debris we had ever seen come off the vehicle and the largest piece of debris we had seen impact the orbiter."

Page picked up the phone and called Wayne Hale. "We've had a major strike on the vehicle," Page informed him. "More details to follow."

When Atlantis was hit by foam three months earlier, the exact source of the debris was murky but the point of impact was clear. The opposite was true in Columbia's case. The team narrowed the origin of the debris to the general area of the tank's left bipod ramp—the same structure that had broken apart the previous October—within the first 10 minutes. But exactly where the briefcase-sized piece of foam had struck Columbia's left wing remained a mystery. The Cocoa Beach camera offered the best angle, but the film was too blurry to provide any details. The Cape Canaveral Air Force Station view was clearer, but the camera angle showed the top of the wing and the impact was somewhere underneath. A third camera much farther south at Patrick Air Force Base was too far away to be of any help.

The film reviewers quickly came to the unavoidable conclusion that they could not tell whether Columbia had suffered serious harm. The breathtaking scale of the impact and the fact it had hit the orbiter all but guaranteed that a team of engineers would be convened to assess the issue. That team, however, would be operating in the dark.

Page's thoughts drifted back to Atlantis' brush with disaster three months earlier. He made up his mind to do something. He knew it was possible to get pictures of the shuttle in orbit using high-powered, top-secret military telescopes and spy satellites. Although Page had never seen such classified photos, he was aware they had been taken on previous missions, such as during the October 1998 flight of Discovery, when a panel fell off the orbiter during launch. After the team finished reviewing films that afternoon, Page told Oliu he was going to ask Hale about getting similar images of Columbia. Hale held a Top Secret clearance and knew how to obtain the pictures from his days as a flight director at Mission Control. Oliu agreed.

"This was absolutely a formal request of these pictures," Page said later. "We could not tell from the film that we had if there was or was not damage to the vehicle. The resolution was not there."

Page went to Hale's office and asked if it was possible to take pictures of Columbia in orbit. He was met with silence—talk about classified capabilities was off-limits. Page rephrased his request: "We can't tell whether or not there was damage to Columbia, and we need pictures of the vehicle in orbit." Hale agreed to look into the possibility.

"I hemmed and hawed when Bob said that," Hale later recalled. "It's got some complications."

N. Wayne Hale, Jr., was two weeks away from officially beginning a one-year stint on February 1 as Kennedy's launch integration manager. The important assignment made him the person responsible for overseeing launch preparations and giving the final "go" for liftoff. The job had usually belonged to astronauts in the past, but Hale's stellar record as a flight director at Johnson had earned him the position. Even so, shuttle managers had to do some arm-twisting before Hale would accept the assignment.

"I never ever wanted to do anything else other than be a flight director," Hale said. "Ron Dittemore had talked to me several times about different things in the shuttle program, and other folks had talked to me about other upper-level management opportunities. A, they can't pay me more money, and B, nothing looked like it was as much fun."

Hale came to Johnson as a flight controller in 1978 after earning a master's degree in mechanical engineering from Purdue University. A decade later, when the shuttle resumed launches following the Challenger accident, he was entrusted with the critical job of flight director for liftoff and re-entry. By 2003, he had supervised launches and landings during a record 28 missions. Overall, he had overseen parts of 40 flights, another record. His idols were Gene Kranz and Chris Kraft, the legendary NASA flight directors of the 1960s. Colleagues at Johnson considered the steady, soft-spoken engineer to be cut from the same cloth as his heroes.

Page's photo request took Hale by surprise. In addition, Page asked Hale to relay the seriousness of the foam strike up the chain of command. Immediately after Page left, Hale called Linda Ham, the chairwoman of the Mission Management Team, to let her know. Ham's office on the fifth floor of Johnson's Building 1 was next door to Dittemore's. Soon, Hale was talking to both of them over Ham's speakerphone. Hale passed along Page's concern that the pictures weren't clear and said an initial photo report with estimates on the debris' size, speed, and impact location would be out by the end of the day. He did not mention Page's request for photos.

"I did not pass on Bob's little comment about 'Gee, are there some other ways to get pictures?' " Hale recalled later. "Remember, the pictures are going to be on orbit. It's not what happened during ascent. So I did not pass that along to them at that time. Whether that was a major screwup on my part—I guess that maybe it was."

Hale left the office that afternoon to catch a plane back to his home in Houston for the three-day weekend; the Martin Luther King Day federal holiday was the following Monday. In his new job, Hale wasn't a member of the Mission Management Team. The management team wouldn't meet again until Tuesday anyway, despite a long-ignored requirement on the books that they convene daily. Hale and most other shuttle managers would all but forget about Columbia's foam strike until then.

Shortly after seeing images of the debris strike that morning, film reviewers at Kennedy, Johnson and Marshall began spreading the word.

Carlos Ortiz, a Boeing systems integration engineer at Johnson, was one of several debris team analysts who periodically traveled to Kennedy for fall and winter launches, when cooler weather increased the chances of ice forming on the tank. He watched the foam strike in Kennedy's film lab that Friday with Page, Oliu and others. After seeing the impact, he called Darwin Moon, a Boeing structure and debris analyst in Houston, with preliminary estimates on the foam's density and composition. It appeared to strike heat tiles on the orbiter's belly near a crook in the leading edge, Ortiz said. Moon called Boeing engineers in Huntington Beach, California, to see if there ever had been an impact like this on the shuttle before. The answer was no.

The film team at Johnson notified the shuttle's Mission Evaluation Room that there had been a potentially serious strike on Columbia. Nicknamed the MER, the Mission Evaluation Room was an engineering control center in Johnson's Building 30. While flight controllers a floor above in Mission Control planned and executed daily activities in space, MER engineers monitored the orbiter's systems at some 40 computer consoles in three narrow, crowded, interconnected rooms.

An entry in the MER's daily log just before 11 a.m. on Friday contained the first reference to the impact:

> [The debris] travels down the left side and hits the left wing leading edge near the fuselage. The launch video review team at KSC thinks that the vehicle may have been damaged by the impact.

The entry noted that two United Space Alliance employees—deputy orbiter engineering manager Bill Reeves and thermal protection system technical manager Mike Stoner—had been notified. United Space Alliance, a Lockheed Martin–Boeing partnership, held the shuttle program's prime operations contract and would lead any analysis of the foam strike.

Opinions on the seriousness of the impact were already forming by that afternoon. Stoner sent out an e-mail to other company managers a few minutes after 4 p.m. He had just spoken with Calvin Schomburg, a NASA tile expert, and Mike Gordon, a subsystem manager for the reinforced carbon-carbon, or RCC, panels that protect the leading edge of the wings from 3,000-degree temperatures during re-entry. Stoner's message downplayed any potential danger:

> Basically, the RCC is extremely resilient to impact type damage. The piece of debris (most likely foam/ice) looked like it most likely impacted the WLE [wing leading edge] RCC and broke apart. It didn't look like a big enough piece to pose any serious threat to the system. . . . At T+81 seconds, the piece wouldn't have had enough energy to create large damage to the RCC WLE system.

The e-mail noted that past analyses showed Columbia should be able to make it back safely even if the impact had penetrated the orbiter's thermal protection system. The message continued:

> As far as the tiles go in the wing leading edge area, they are thicker than required and can handle an area of shallow damage which is what this event most likely would have caused. They have impact data that says the structure would get slightly hotter but still be OK.

Meanwhile, rumors were rampant at Kennedy that the shuttle might be landing early on the following Tuesday. It was standard procedure to bring the shuttle home if major problems developed after launch that posed an immediate threat. An early return wouldn't help if there was damage to Columbia's thermal protection system, but still the rumors were flying. Page phoned MER manager Don McCormack, who assured him the rumor wasn't true. McCormack noted the call in the log. Before leaving that night, McCormack made a final entry on the strike as part of the handover from his shift to the next.

> Debris impact on port wing edge—appears to have originated at the ET [external tank] fwd bipod—foam? If so, shouldn't be a problem.

By late Friday, United Space Alliance officials started organizing the debris assessment team that would begin preliminary work that weekend. The effort would be chaired by Pam Madera, a United Space Alliance subsystem manager for vehicle and systems analysis. Because the huge foam strike had been classified as "out of family," NASA participation in the effort was required. Joyce Seriale-Grush, a NASA engineering manager, called Rodney Rocha, the space agency's division chief for shuttle structural engineering at Johnson. She told him to work with the contractors and coordinate any government resources the team might need.

Rocha was a veteran NASA engineer with a reputation for being smart, conscientious and meticulous—a pro at what was referred to in NASA-speak as "pounding a problem flat." Some managers also considered him opinionated, stubborn and occasionally confrontational. The 51-year-old Texan had grown up in Corpus Christi and, by age 13, had helped found the city's first astronomical society. He had dreamed of designing telescopes one day, but as the space race became a national obsession, Rocha began to set his sights on a job with NASA. He graduated from high school in 1969, just weeks before the Apollo 11 moon landing, and attended a junior college to study math.

When recruiters from the University of Texas came by touting their aerospace engineering program and connections with NASA, Rocha was sold. He enrolled in 1971 on an academic scholarship and graduated in 1974. Although he was accepted to graduate school, he interviewed with the rest of his class when NASA and a group of aerospace contractors came to campus. Several companies made offers, but no word came from NASA. Rocha agreed to take a job with McDonnell Douglas. But he worked up the nerve to call NASA—collect—and ask why he hadn't gotten an offer. A NASA representative called back later with an apology. They had lost his application.

Rocha reapplied and was offered a job. He went to work at Johnson in 1974 as a technical intern recruit and was assigned to Building 13, the place where he would spend the next 29 years doing structural engineering. His first assignment was to analyze how solar panels on a Soviet spacecraft would be affected by the vibration of docking with an Apollo capsule during a joint 1975 goodwill mission. He was named chief of the shuttle's structural engineering division a year before Columbia's launch.

Seriale-Grush phoned Rocha in his office late that Friday afternoon.

"She said 'Rodney, did you know that something hit the wing on ascent?' " Rocha recalled. "I gasped. I said, 'Really? Tell me more about it.' She told me it looked like the bipod foam ramp and that it had happened on STS-112 [Atlantis' flight], too.' "

Rocha immediately went down the hall to query thermal engineers on how serious that sort of impact could be. Very serious, they said. He began to think about possible ways to change how Columbia flew during re-entry to minimize heating if parts of the orbiter were damaged. Meanwhile, Oliu e-mailed engineers the results of Kennedy's Day One film review. The report said a large object had broken off the left bipod area and hit underneath the left wing 81 seconds after launch. It contained a clip of the Cape Canaveral Air Force Station footage as well as still frames from the same video. Later that night, Rocha saw the images for the first time in his suburban Clear Lake, Texas, home. He replayed the clip over and over in his living room for his wife, Robyn.

"I was astonished," Rocha said. "I had never seen anything like that before—ever. The debris cloud was particularly intriguing. Am I looking at the debris shattering completely, or is that part of the wing's TPS [thermal protection system] with it?"

Boeing engineers went to work Saturday to begin the process of answering four critical questions: Where exactly did the foam strike Columbia? What was the possible extent of the damage to the orbiter's thermal protection system? How would the damage affect heating of the orbiter as it re-entered Earth's atmosphere? How well would the ship hold up under those changes in heating?

By Saturday morning, Ortiz already had come up with preliminary estimates on the location and severity of the impact—literally on the back of an envelope in his hotel room, another engineer recalled. The foam chunk appeared to be as large as 20 by 16 by 6 inches. It was believed to have hit at a relative speed of 510 mph and at an angle of up to 20 degrees, which raised the possibility of a more direct strike instead of a glancing blow. Using that information, Dennis Chao, a Boeing engineer with 25 years' experience in thermal analysis, came up with a possible impact location on Columbia's belly. The best guess was an area on the black heat tiles between the left wing's leading edge and the landing gear door a few feet away.

The estimate also put the strike dangerously close to one of the 22 U-shaped reinforced carbon-carbon panels that lined the leading edge. As Stoner had noted in his e-mail the previous day, there was a widespread belief that the RCC was impervious to foam damage. Only twice in 113 missions—during a 1991 flight of Atlantis and a 1999 flight of Discovery—had debris impacts damaged the RCC during flight. The causes never were explained, but neither incident was considered serious. Even so, after Ortiz spoke with Moon that Saturday afternoon, the engineers' preliminary number-crunching came up with some disturbing results.

A computer model dubbed "Crater" estimated the damage to Columbia's tiles. Engineers had developed Crater during the Apollo program in 1966 to predict the threat to spacecraft from tiny rock fragments in space. The formula had been modified during the early years of the shuttle program to predict damage from foam, ice and other tank debris. Data for the model were based on impact tests in 1979 and 1999, when small cylinders of foam not much bigger than cigarette butts were fired at single blocks of tile. No tests were done with larger pieces of foam, like the briefcase-sized chunk that hit Columbia. Nor had the tests fired objects at larger groups of tiles installed together as they are on the shuttle. Crater had been used most often on the day of launch to predict possible debris damage if an inspection team spotted ice forming on the tank before liftoff. The formula never had been used during a shuttle flight to estimate tile damage from foam.

Crater analyses, like most of Boeing's other shuttle production and operations tasks, had been done for years at the company's Huntington Beach office. But responsibility for that and many other shuttle operations officially had moved from California to Houston earlier that month in an efficiency and cost-cutting move. As a result, Boeing engineer Paul Parker, who had been trained on Crater but had used the program only twice before, was assigned to help perform the analysis. Despite his inexperience, he knew enough about Crater to have concerns. The foam block that hit Columbia was estimated to have a volume of 1,200 cubic inches. That was at least 400 times greater than the largest foam cylinders used in impact tests to develop Crater.

When all of the estimates on the foam chunk's size, trajectory and impact site were loaded into Crater, the model calculated that the strike would have penetrated completely through Columbia's heat tiles. That meant the wing's vulnerable aluminum airframe could be exposed to blowtorch-like gases during re-entry. While Crater couldn't be used to calculate RCC damage, engineers had a similar algorithm that could. The formula was based on 1984 tests that shot cigar-sized ice projectiles at RCC panels. Calculations showed Columbia's RCC needed to be at least 0.246 inches thick to avoid penetration from the foam strike. The panels were only 0.233 inches thick.

The uncertainty made it more critical than ever that engineers pinpoint the precise location of where the debris hit. Additional photo analyses on Saturday, however, shed little new light on the mystery. Boeing engineers were rapidly coming to the realization that it would be impossible to have any confidence in their assessment without photos of Columbia in orbit.

Independently, Rocha already had reached a similar conclusion. He typed an e-mail late Sunday raising the possibility of having Columbia's astronauts survey the top of the wing through the hatch porthole on the ship's mid-deck. Rocha sent the message after midnight to shuttle engineering manager Paul Shack and others.

There was no reply.

Several members of the debris assessment team met informally on Monday's Martin Luther King Day holiday to review preliminary work done over the weekend. During the next three days, the group would grow to more than three dozen people from NASA and three contractors—United Space Alliance, Boeing and Science Applications International Corporation. Chairwoman Madera led the meeting.

The team discussed Ortiz's work at narrowing down the size and trajectory of the foam. Boeing's Huntington Beach office faxed Parker a copy of the 1999 foam impact tests done by Southwest Research Institute in San Antonio. It confirmed that the piece of foam that hit Columbia was vastly larger than any used in the tests. There was little new photo information. A comprehensive report from the Intercenter Photo Working Group that typically comes out four days after launch had been delayed while Page's computer was replaced. Engineers were more concerned than ever, and a new worry had begun to develop. The likely impact zone included part of the main landing gear door on Columbia's left wing. If the strike had breached the door, super-hot gases would blow in during re-entry, destroying the landing gear's tires—or worse.

Team members unanimously agreed that images of Columbia were needed to more precisely locate the damage. The consensus was to seek pictures from one of the high-powered ground telescopes the U.S. military uses to gather intelligence. That desire was documented in an e-mail sent out by Madera later that Monday outlining plans for the group's first official meeting on Tuesday afternoon.

On Tuesday morning, officials convened the flight's second Mission Management Team meeting. Shuttle managers, some of whom were returning from a three-day weekend, had not met since the day after Columbia's launch, before the foam strike had been discovered by film reviewers.

The management team's job was to oversee shuttle missions during flight and deal with any serious issues. The group met in the so-called Action Center, a cavernous gray conference room on the second floor of Johnson's Building 30, the same facility that housed Mission Control and the MER. Rows of logos celebrating past NASA missions adorned the walls. The center of the room was occupied by a long black conference table with six black chairs on each side and one at each end. Rows of blue chairs surrounded the table on three sides, giving the room a theaterlike feel. The job of chairing the meetings usually fell to NASA's shuttle program integration manager, the official responsible for making sure each flight's overall mission requirements were met. During Columbia's flight, that was Linda Ham, who sat at the head of the table.

Ham was a no-nonsense, take-charge engineer whose smarts and ability to think on her feet had propelled her career upward in the shuttle world at Johnson. The 42-year-old working mother had grown up as Wisconsin math whiz Linda J. Hautzinger in the town of Kenosha. After graduating from high school, she enrolled in the nearby University of Wisconsin, Parkside. She lived at home, commuting to school and working to pay her way through college by coaching high school and junior high gymnastics in her spare time. Despite the heavy load, she had almost completed a degree in mathematics by the end of her second year. She thought about earning a Ph.D. and starting a career in teaching.

Ham graduated in 1982 with bachelor's degrees in both mathematics and applied science. She was 21 years old but had never flown on an airplane. In fact, she had never traveled outside the state of Wisconsin. She interviewed for several local jobs and was on the verge of accepting an offer. Then, a few weeks before graduation, she decided, on a whim, to send her résumé to several places she always had dreamed of working. One of them was NASA. A section head from Mission Control called and virtually offered her a job over the telephone. She visited Houston, accepted a position as a flight controller monitoring the shuttle's onboard propulsion systems, and never looked back.

Ham thrived in the virtually all-male bastion of Mission Control and quickly became a trailblazer. After six years as a flight controller, she was named the section head of her division, becoming the first woman promoted to that job. She met her husband, Ken, in 1987 when he was a Navy flier assigned to Johnson's astronaut office as a crew member aboard the "Vomit Comet," a plane used to simulate weightlessness. The two married in 1990, the same year Ham's husband got his wings as a Navy fighter pilot. They continued to live apart afterward for 10 months of the year as the Navy transferred him to new assignments. Ham embarked on a new assignment of her own in 1991 as NASA's first female flight director.

By 1994, the couple had two sons. Like many Navy wives, Ham led the life of a single mother as her husband drew assignments in the Mediterranean flying combat missions over Iraq and Bosnia. A fitness freak, she could be seen dutifully pushing her double baby stroller to the gym after long days at the office. Being in shape also was essential to her new career goal: astronaut. To improve her chances, she went back to school and earned a master's degree in aerospace engineering from the Naval Postgraduate School in 1997. Her poor eyesight, however, was a problem. She had laser surgery to improve her vision. But NASA wasn't accepting applicants who had undergone the procedure. Meanwhile, Ham's husband applied for the astronaut corps and was accepted in June 1998. Ham wouldn't be an astronaut herself, but there would be one in the family.

The following year, Ham became the first woman to serve as flight director for the critical liftoff and landing phases of a shuttle mission. Dittemore saw her potential and selected Ham as his technical assistant in 2000. Less than a year later, she was promoted to the key job of program integration manager. Some at Johnson quietly wondered whether her meteoric advance had been too far, too fast. But when Columbia launched on Jan. 16, Ham was a force to be reckoned with.

An hour before the management team meeting, Boeing engineers had briefed MER manager McCormack and other NASA engineers on what the debris assessment group had learned so far. Toward the end of the meeting, McCormack relayed that information to Ham and the rest of the teleconference.

"As everyone knows, we took a hit on the—somewhere on the left wing leading edge and the photo-TV guys have completed, I think, pretty much their work, although I am sure they're still reviewing their stuff," McCormack said. "They've given us an approximate size for the debris and an approximate area where it came from and approximately where it hit. . . . We're also talking about what you can do in the event we have some damage there."

Like other senior managers, Ham wondered how the foam issue might affect future missions. They were confident the foam strike posed little threat to Columbia. But the issue would have to be dealt with before the next launch. When Columbia landed, flights to build the International Space Station would resume. Shuttle and station program managers at Johnson were under withering pressure from NASA headquarters in Washington to make sure the first phase of assembly was finished by Feb. 19, 2004. Senior agency officials had drawn a line in the sand that the core of the outpost be completed by that date. The project had a history of consistently falling behind schedule and going over budget. Ham was keenly aware of this. In the convoluted world of Johnson's shuttle bureaucracy, the real threat wasn't to Columbia, it was to the flimsy rationale adopted the previous October to launch Endeavour. That same rationale or a new one would be needed to keep station missions flying after another foam incident during Columbia's launch. Ham voiced this concern by recounting Atlantis' foam strike and the Flight Readiness Review for Endeavour that followed. She told the teleconference: "I'm not sure that the area is exactly the same where the foam came from, but the [aerodynamic] properties and density of the foam wouldn't do any damage. . . . I hope we had good flight rationale then."

McCormack recalled a 1997 Columbia mission when there was damage on the tiles between the wing's leading edge and the main landing gear door. He said he would gather data from the incident.

"I really don't think there is much we can do. So it's not really a factor during the flight because there is not much we can do about it," Ham concluded. "But what I'm really interested in is making sure our flight rationale to go was good, and maybe this is foam from a different area, and I'm not sure and it may not be co-related, but can you try to see what we have?"

Later in the meeting, shuttle manager Lambert Austin reported the foam might have broken off of the same bipod ramp where Atlantis' tank had lost foam. Engineers were looking at whether any options were available to change the shuttle's re-entry if needed.

When the meeting was over, Ham went back to her office and e-mailed Dittemore about the foam rationale that tank manager Jerry Smelser had presented two months earlier at Endeavour's Flight Readiness Review. Dittemore was out of town that week, attending the test firing of a shuttle booster at the manufacturer, ATK Thiokol, in Utah. He replied a few minutes later:

> You remember the briefing! Jerry did it and had to go out and say that the hazard report had not changed and that the risk had not changed. . . . But it is worth looking at again.

Ham replied an hour later.

> Yes, I remember. It was not good. I told Jerry to address it at the ORR [Orbiter Rollout Review] next Tuesday (even though he won't have any more data and it really doesn't impact orbiter roll to the VAB). I just want him to be thinking hard about this now, not wait until IFA review to get a formal action.

Ham was thinking about Atlantis' pending move from its hangar to the Vehicle Assembly Building at Kennedy, where it would be connected to its tank and boosters for the next space station flight. Despite the two foam incidents, Smelser's presentation at Endeavour's Flight Readiness Review could provide shuttle managers with a rationale for rolling Atlantis to the launchpad in another week and a half—if the rationale was solid. Ham and other managers, however, quickly concluded that it wasn't.

A few minutes after Dittemore's message, McCormack e-mailed Ham the briefing charts from Smelser's presentation. After looking at Smelser's rationale—nothing was different so everything should be OK—McCormack shipped the charts with a note: "FYI—it kinda says that it will probably be all right."

Ham summed up her unhappiness with the rationale in another e-mail to Dittemore a few minutes later:

> The ET rationale for flight for the STS-112 [Atlantis] loss of foam was lousy. Rationale states we haven't changed anything, we haven't experienced any 'safety of flight' damage in 112 flights, risk of bipod ramp TPS [foam] is same as previous flights. . . . So ET is safe to fly with no added risk.
>
> Rationale was lousy then and still is. . . .

Others at NASA weighed in on the issue as well.

Shack discussed the subject in an e-mail that morning to Rocha and other shuttle engineers. He mentioned the tank project's presentation before Endeavour's launch and noted that the charts said foam "loss over the life of the Shuttle program has never been a 'Safety of Flight' issue. They were severely wire brushed over this and Bryan

O'Connor [associate administrator for safety] asked for a hazard assessment for loss of foam."

The wire brushing referred to O'Connor's debate with Smelser on whether the foam strikes officially should be classified as a "safety of flight issue" or an "accepted risk." Like most others who heard Smelser's argument, Shack had left unimpressed.

"A lot of us walked out of the FRR and kind of looked at each other and said 'That stinks,' " Shack recalled later. "The general attitude was that it didn't smell right."

Now, as the semantics debate raged anew at Johnson, shuttle managers still failed to grasp the potential seriousness of foam strikes. Columbia, they thought, would be fine. Managers knew however, that something would have to be done to pave the way for launching Atlantis after two major foam strikes during the last three missions.

"After this flight when it [the foam ramp] came off, we knew we had a significant issue," Ham later said. "We were going to have to change."

The debris assessment team officially convened for the first time on Tuesday afternoon in a ground-floor conference room in Johnson's Building 13. About a dozen people gathered around a table in the large blue conference room, while others joined via teleconference. Rocha darkened half the room and played a film loop of the debris strike on a projector. The loop repeated over and over. Occasionally, the discussion would all but stop as engineers stared mesmerized at the screen.

Much of the meeting focused on the possible limitations of Crater and whether it could accurately predict the level of damage to the tiles—assuming the foam had even hit the tiles. There was considerable discussion on the size of the debris. Ortiz briefed the group on his analysis of the impact's location and severity.

"It seemed to be the same old discussion," Rocha said. "Where is it hitting? What is that cloud? How big is that piece? Is it really coming from the bipod area or not? Is it spinning? Can we see damage?"

A sense of urgency filled the room. Initial analyses had shown the damage was potentially serious. Those analyses, in turn, would help determine how far the group went in exploring possible changes to Columbia's re-entry. It was clearer than ever that pictures of Columbia in orbit were crucial. Rocha volunteered to take the photo request to managers in Johnson's engineering directorate. He went back to his office and e-mailed a request to Shack and two other high-level engineers a few minutes later. Rocha wrote:

> We will always have big uncertainties in any transport/trajectory analyses until we get better, clearer photos of the wing and body underside. Without better images it will be difficult to even bound the problem and initialize thermal, trajectory, and structural analyses. Their answers may have a wide spread ranging from acceptable to non-acceptable to horrible, and no way to reduce uncertainty. Thus, giving MOD [Mission Operations Directorate] options for entry will be very difficult. Can we petition (beg) for outside agency assistance? We are asking for [Johnson engineering director] Frank Benz with Ralph Roe or Ron Dittemore to ask for such.

The e-mail concluded that "despite some naysayers" there were, in fact, some options to discuss for changing Columbia's re-entry if the ship was determined to have damage.

"I sensed that time was running out," Rocha said later. "We needed to tell the mission operations folks something on the situation if we were going to fly differently. This was Tuesday. We've been given the action to report by Friday morning at 8 a.m. to Linda Ham. . . . We needed an answer right then. We needed a photo just to start this analysis correctly."

By early Wednesday, three requests for photos of Columbia in orbit had begun to independently percolate up the NASA chain of command. Besides Rocha's request, United Space Alliance shuttle manager Bob White had contacted Lambert Austin on Tuesday afternoon after hearing from several of his company's engineers on the debris assessment team. Austin then left a phone message for Hale before calling Air Force Lt. Col. Timothy Lee at Johnson's Defense Department support office to ask about getting the process started.

In Florida, Hale arrived at the office early Wednesday. After listening to Austin's message, he went to work on the photo request Page had made five days earlier. Hale spoke with Dave Phillips, a NASA employee who worked as a liaison with the 45th Space Wing at nearby Patrick Air Force Base. Hale asked Phillips to look into the photos and Phillips agreed. Phillips walked across the hall to talk with an Air Force officer about the request and was surprised by his response.

"I asked, 'Can you?' " Phillips recalled, "and they said 'We have already been asked.' "

Austin's request had gotten there first. Now, with two queries for photos of possible damage to the shuttle, the Air Force was eager to help. Military officials in Florida contacted the U.S. Strategic Command at Cheyenne Mountain, Colorado, a nerve center for space surveillance. NASA had a memorandum of understanding for such requests with the National Imagery and Mapping Agency, a federal organization that supports U.S. military and intelligence needs.

"We got turned on," Phillips said. "We went up to the right places and got affirmative answers that 'Yes, we will do it.' "

The request remained turned on for only an hour and a half.

Immediately after talking with Phillips, Hale called Phil Engelauf, who represented Johnson's Mission Operations Directorate, to let him know about the call and ask that Mission Control make a formal request through official channels. Meanwhile, Austin called Ham to fill her in. Austin knew he should have gotten Ham's approval first—but he hadn't. Ham made it clear Austin did not have the authority to request photos and asked who needed them. Austin replied that Bob White and others at United Space Alliance had wanted the pictures.

Ham immediately began calling around. She phoned Roe's vehicle engineering office to see if it wanted photos. Roe, after checking with the MER, said his office had no need. His judgment was reinforced a few minutes later during a conversation with Schomburg, the tile expert.

"Calvin's discussion with me centered around the fact that the analysis that was being performed—and he had seen preliminary runs—was a worst-case analysis, meaning they assumed that the tile was gone completely down to the densified layer, which is just a very thin layer of tile remaining attached to the structure of the orbiter," Roe said. "The conclusion that we came to in that discussion was, well, we are doing a worst-case analysis. We will let that analysis tell us what worst case is, and information from a photo . . . we didn't think would improve the analysis."

Ham rapidly was reaching the same conclusion.

"Wing leading edge, of course, that didn't really even cross my mind, I don't think, at the time," Ham recalled. "So Ralph saw no reason for me to pursue the imagery. . . . So I asked the MER, and the MER says, 'Well, no, we don't know of anything. We will go check too.' So, I hung up the phone, and they will call me back. Well, they check . . . and 'No, we're not asking.' Then, they check with Ralph, and say, 'No, he's not asking.' So then they call me back and say, 'Well, nobody we know is asking.' I said, 'All right.' . . . Then I thought, Loren Shriver, USA [United Space Alliance], and I called him. He searched around, and he said he couldn't find out who was looking. So I called Wayne back."

Shriver found out during subsequent calls that the request had come from Page and White. By then, however, it was too late. Dittemore and Ham already had agreed that the request should be turned off. Ham phoned Hale about 90 minutes after his first call to Phillips to give him the news.

Months later, Hale recalled the conversation this way:

"She said, 'Now Wayne, we don't have anybody who wants a requirement. You know this is a really busy flight. . . . We've got all these science payloads and it's carefully integrated and you've got to fly pointing this direction for this payload. . . . We don't know what we want to take pictures of. Every time we've taken these images before they have never been useful to us.' . . . And then she had the clincher argument. She said, 'Wayne, you know, even if there was damage, there's nothing we can do about it.' And it wasn't a fatalistic 'We're going to die.' It was just that OK, so you come back and say that we have got damage, we don't have a tile repair kit. You know, all this stuff about flying different entry profiles which everybody has talked about ad nauseam there—there isn't anything you can do about it. You just have to hope it holds together. So there we were. And I had to kind of nod my head and say,

'Well, you know Linda, you're right.' She said, 'So, you know, well let's just turn this off' or words to that effect. And I said 'OK.' "

Nevertheless, when Hale hung up the phone, he was angry. He got up and walked down the hall. He thought about delaying his call to stop the photo request. Maybe if he waited a day, the pictures already would have been taken. Ultimately, however, Hale followed Ham's directive. He called Phillips and told him to forget about the request he had made on behalf of Page. When Hale broke the news to Page, Page was devastated. Page went back to his office and logged an entry in his personal notes for posterity:

> Linda Ham said it was no longer being pursued since even if we saw something, we couldn't do anything about it. The program didn't want to spend the resources.

Next, Hale called Mission Control to turn off the official request he had made of Engelauf earlier. He spoke with flight director Steve Stich.

"[Stich] then went through this same kind of agony that I did about, well, do I really want to turn it off or do I just want to sit on my hands and let it happen and say I didn't get it turned off in time and take the consequences," Hale said. "But he finally felt like he had to turn it off. So he did."

Stich spoke with Roger Simpson, a NASA official at Peterson Air Force Base in Colorado who acted as a liaison with Cheyenne Mountain. By the time the request was canceled, it had gone all the way up the Air Force chain of command to a four-star general. The military was ready to help. "I told them that we did not require the data on this mission and that they could turn off their system, which was in high gear," Stich noted in an e-mail the following week. "In hindsight, I probably should have let them go since they had worked very hard on the [military] end and they may not respond as well next time since we 'cried wolf.' "

Simpson sent a message on Thursday thanking the Air Force for its support. But he stressed that future requests for assistance like Hale's and Austin's should not be honored unless they were made by proper authorities. Simpson wrote:

> Your quick response in arranging support was exceptional, and we truly appreciate the effort and apologize for any inconvenience the cancellation of the request might have caused. Let me assure you that, as of yesterday afternoon, the shuttle was in excellent shape. . . . One of the primary purposes for this chain is to make sure requests like this one do not slip through the system and spin the community up about potential problems that have not been vetted through proper channels.

Incredibly, almost all of the shuttle officials who discussed getting photos for Page and United Space Alliance managers on Wednesday morning later said they were oblivious to the fact that Rocha had made a similar request on behalf of the debris assessment team. While the debate went back and forth on Hale and Austin's attempts, Shack had visited Roe's office to discuss Rocha's request. Roe had left to go to Utah for the same booster test that Dittemore was attending. So Shack spoke instead with Roe's deputy, Trish Petete.

While Shack and Rocha usually had a cordial working relationship, the two didn't always see eye to eye. The previous November, Endeavour's launch had been delayed while engineers studied whether an access platform that accidentally bumped the shuttle's robot arm had caused damage. Shuttle managers wanted an expedited engineering analysis on the structural forces the arm would be subjected to during launch and landing without delving into the time-consuming minutiae of scenarios they considered irrelevant. Rocha felt differently.

"We still needed to assess them," Rocha said. "I got a little steamed. I said 'Don't tell me what to do on this. I know what to do. We know what to do. I know what he wants, but we are going to look at it anyway and we are the ones who are going to decide what it is we should look at and what we shouldn't look at.' "

It wasn't the first time the two had disagreed. Similar exchanges had earned Rocha a reputation among some as a maverick. Shack saw him as a mercurial personality who occasionally had cried wolf in the past. He presented the photo request to Petete as having come from Rocha individually—not from a team of engineers working to assess possible damage to Columbia. Shack added that his "Rodney filter" was turned on.

"I said, 'What do you mean by that?' " Petete recalled. "He said Rodney had a reputation like Chicken Little—you know, the sky is falling all of the time. . . . I think that influenced the way that Paul approached me, at least with Rodney's concerns."

The two decided the best thing to do was wait until the damage assessment team came back with their findings, never considering that the accuracy of the team's findings could depend on the photos. If the results looked bad, then images of Columbia could be requested later. A few minutes later, Petete stopped Schomburg in the hall to ask his opinion.

"I said, 'Well, I don't necessarily recommend that we take that photo now. I think you all's recommendation of doing the analysis, and seeing what the analysis says, would be the proper thing to do,' " Schomburg remembered.

An entry in the Mission Evaluation Room's log at 10:37 a.m. Wednesday had documented Ham's decision to cut off the photo request made by Hale. A little more than an hour later, Shack referred to the entry in an e-mail he sent to Rocha and others:

> FYI—According to the MER, Ralph Roe has told the program that Orbiter [engineering] is not requesting any outside imaging help.

It would be weeks before Rocha and the rest of the team learned the MER entry referred to a photo request that wasn't theirs.

"By 11:45 a.m.," Rocha said, "I got the Paul Shack reply. . . . It instantly stunned and astonished me. Questions spun in my head. What could it really mean? What happened to our plea? How far did it go in the management chain? Why was our photo request denied? Did the space shuttle program know something already about any wing damage that we did not and, if so, had determined that everything was safe? Was the photo too expensive for NASA or too troublesome to arrange via an outside agency? Since I assumed Shack had surely provided adequate explanation of our debris assessment team's urgent photo need, and conveyed our request to Ralph Roe or another space shuttle program manager, then I interpreted this remark from Ralph Roe as a space shuttle program hard position and as a negative reply made expressly to our team request. . . . It seemed too incredible and wrong that an integrated technical team of experts charged officially with making the assessment by January 24 was being told no."

In fact, shuttle managers had said no to the requests from Page and United Space Alliance. Most weren't aware of Rocha's request. Roe hadn't even been told.

Rocha had assumed that, if nothing else, the team's need for a photo had been conveyed to the top in responses to an e-mail Ham had sent out earlier Wednesday morning.

"Can we say," Ham asked, "that for any ET [external tank] foam lost, no 'safety of flight' issue can occur to the Orbiter because of the density?"

Ham was looking for reassurance on the conventional wisdom that had evolved among shuttle managers: Foam wasn't dense enough to do serious damage. NASA's *Integrated Hazard Report* 37 clearly stated, however, that lost foam could result in the catastrophic loss of the shuttle and its crew. But if shuttle managers could prove differently, not only would that mean Columbia was safe, but it would provide the badly needed rationale to keep the next missions on schedule. Dittemore replied to Ham a few minutes later: "Another thought, we need to make sure that the density of the ET foam cannot damage the tile to where it is an impact to the orbiter."

Ham's question was forwarded down to a number of engineers, including Rocha. He sent back an emphatic no and reiterated the team's need for a photo. Eventually, the collected responses were sent to Ham in an e-mail from Austin that afternoon:

> NO. Recall this issue has been discussed from time to time since the inception of the basic "no debris" requirement. . . . It is not possible to PRECLUDE a potential catastrophic event as a result of debris impact damage. . . . Foam loss can result in impact damage that under subsequent entry environments can lead to loss of structural integrity of the orbiter area impacted. . . . The most critical orbiter bottom acreage areas are the wing spar, main landing gear door seal and RCC panels.

Knowing that earlier photo requests had been denied, Austin made no mention in his e-mail of the debris assessment team's need for images of Columbia. When the group met that Wednesday afternoon, many of the engineers already had heard that a photo request—theirs, they assumed—had been turned down. The rest were stunned by the news. Time was running out.

Again, the team ran through the latest calculations: The size of the impact. Possible changes to re-entry. The likely extent of the damage. The discussion, however, kept returning to the pivotal issue of where the foam had hit. There was a growing consensus that tile damage on the orbiter's belly was the most likely scenario, although concern continued to grow about the seal on the landing gear door. Photos of Columbia in orbit might have pinpointed the location and revealed the extent of the damage. But now, there would be none. When Rocha returned from the meeting, he typed out a scathing e-mail to Shack and other shuttle engineering managers ripping the program's decision not to seek pictures.

"In my humble technical opinion," the e-mail draft said, "this is the wrong (and bordering on irresponsible) answer from the SSP [space shuttle program] and Orbiter *not* to request additional imaging help from any outside source. I must emphasize (again) that severe enough damage . . . could present potentially grave hazards. The engineering team will admit it might not achieve definitive high confidence answers even with additional images, but, without action to request help to clarify the damage visually, we will guarantee it will not. Can we talk to Frank Benz before Friday's MMT [Mission Management Team] meeting? Remember the NASA safety posters everywhere around site stating 'If it's not safe, say so?' Yes, it's that serious."

When Rocha finished typing the message, he debated for several minutes whether to send it.

"I hovered over that send key, but what was working through me was 'Well, maybe I better work through the line of command,' " he said later. "I may get in trouble for this. I don't know yet. I don't have any facts yet. I thought it was a pretty strong, biting message, and I guess I wanted the system to work it a little bit more. I was angry and still upset."

Rocha didn't press the send key. Instead, he sent a far more subdued message to Shack a few minutes after 6 p.m., asking about the "no" from Roe referred to in the MER log:

> Paul, Can you tell us more on Roe's negative answer? Is he and the SSP waiting on our analysis results first (Friday to the MMT) or what? What is Frank's [Benz's] position?

There was no reply. As Rocha sat in his office stewing, he decided to pick up the phone and call Shack directly.

Shack answered, put Rocha on speakerphone, and asked Seriale-Grush to join the discussion. Rocha wanted to know if Benz had been told of the photo request. Shack said no. The conversation began to get heated. Rocha asked Shack to set up a meeting with Benz. By this time, Rocha was shouting.

"He [Shack] said, 'If you want a meeting with Frank, you can go ahead on your own and arrange one with Frank. I am not going to be Chicken Little about this,' " Rocha remembered. "And that is when I shouted back, 'Chicken Little? Paul, the program is acting like an ostrich with its head in the sand.' "

Seriale-Grush broke the tension by joking that Chicken Little and an ostrich both were birds. But one more opportunity to bring the team's request to the attention of senior management had hit a dead end.

"I thought Rodney was being over-emotional," Shack explained later. "Generally, when an engineer comes to me with an emotional argument, I try to bring it down to technical terms because I know an emotional argument usually doesn't win your case for you. If you go to a higher-level manager and you are basing your argument on emotion rather than data, you are going to lose."

Rocha, however, wasn't done. He was ready to bang on shuttle managers' doors in Building 1 to get his point across. He went around his division offices to find supporters.

"After my shouting match with Paul," Rocha said, "I was fighting mad. I decided I would go to my division management for guidance and to discuss my next course of action. I also went to rally their support. I am ready to break the doors down. I want to go up to Frank Benz or Ron Dittemore or Ralph Roe—somebody like that in Building 1—and I want someone to go with me."

It was late in the day, however, and most people, including his immediate supervisor, had already left. Finally, he found Doug Drewry, another NASA engineering manager who was a deputy division chief. Rocha was waving around a copy of his unsent e-mail, threatening to ship it off to top shuttle managers. Drewry asked that he reconsider and suggested that they wait and see if the assessment team's analysis showed the potential for disaster. If so, they would try to get the photo request turned back on.

"After hearing that, I felt deflated," Rocha said. "I ran out of steam. I didn't verbally say this to myself, but I unconsciously felt no level of engineering manager is going to support this photo request, not him, no one is sticking their neck out on this. No one is going with me to Building 1. They are not going to, and that proposal he just made—let's make a conservative case, then we will do something if it shows bad—that sounded reasonable."

Rocha went home.

There was one more attempt to get images for the team the next morning, on Thursday.

A United Space Alliance engineer had suggested that Rocha speak with Barbara Conte, a contractor in mission operations who was familiar with the government's capabilities for taking pictures in orbit. Rocha had left her a telephone message on Wednesday night explaining the team's predicament. She returned his call on Thursday morning.

"I went over what had just happened, the 'no' answer, what our team responsibility was," Rocha said. "I was very agitated. I can't understand why they are saying no. 'What I want to ask you, Barbara, is I heard that you have access to photo assets. Is that true?' "

Conte replied that Air Force assets routinely were used to photograph the separation of the tank and boosters from the shuttle. She refused to give Rocha any specifics on the capabilities but later asked around about getting pictures on his behalf. At an unrelated meeting that morning, Conte discussed Rocha's concerns with flight director LeRoy Cain and offered to get the process going. Cain consulted with Engelauf and sent out an e-mail titled "Help with debris hit" shortly after noon:

> The SSP [space shuttle program] was asked directly if they had any interest/desire in requesting resources outside of NASA to view the Orbiter (ref. the wing leading edge debris concern). They said, No. After talking to Phil, I consider it to be a dead issue.

The debris assessment team met for the final time on Thursday afternoon. It would have to finish its work without pictures. Rocha was late because of an earlier, unrelated engineering meeting. When that meeting had ended, Seriale-Grush approached Rocha about the previous night's heated phone call with Shack. Maybe, she suggested, Rocha's concerns would be eased if he spoke with Schomburg, NASA's resident expert on the shuttle's heat tiles.

Schomburg hadn't missed an opportunity to argue in e-mails, meetings, and private discussions with other shuttle managers that the foam strike posed little threat to the tiles or Columbia. His e-mail on Tuesday morning to Shack and Seriale-Grush had helped set the tone for how the strike would be treated: "FYI—TPS took a hit—should not be a problem—status by end of week."

The 59-year-old Texan was one of the tile system's founding fathers. He was an expert's expert, supremely confident in his knowledge, often to the point of being abrasive and outspoken. His name was on about a dozen tile-related patents. Shuttle managers regarded him almost with reverence as the final authority on the subject. He had come to work at Johnson in 1965 as a co-op student at the University of Houston and was hired when he graduated in 1967. Later that year, a launchpad fire during a practice countdown at Cape Canaveral killed three Apollo astronauts. Schomburg played a major role in analyzing the accident and helping to redesign the module. He began working on heat tiles in 1968, and by 1970, was doing it full-time. During the next three decades, he rose through the ranks to become Roe's technical manager for the tile system. Although Schomburg was considered

Johnson's final word on the tile part of the thermal protection system, his expertise didn't extend to the reinforced carbon-carbon surfaces that insulated the parts of the orbiter exposed to the greatest heating. That job fell to Don Curry, Johnson's resident RCC expert.

Rocha, like most of the debris assessment team, suspected the foam had probably struck tiles somewhere on the orbiter's belly near the landing gear door. He went upstairs to Schomburg's sixth-floor office to talk. The two met in a room nearby.

Rocha suggested that the program had to assume Columbia's condition was unsafe until proven otherwise. Schomburg said there was no reason to believe that. History showed it was a tile refurbishment issue—nothing more. Rocha began to talk about possible re-entry alternatives to minimize heating, including landing the shuttle at its backup runway at Edwards Air Force Base in California's Mojave Desert instead of at Kennedy. Schomburg replied the re-entry already planned was the most benign the shuttle could fly. By this time, Rocha said, the two were shouting.

"He had his points," Schomburg remembered. "Rodney was concerned that the vehicle had been damaged. He was looking at all possible alternatives."

The debate ended moments later.

"He leaned forward in my face with his index finger," Rocha recalled, "and said, 'Well, if it is that bad, there is not a damn thing we can do about it.' That just astonished me, and I said, 'Calvin, I grant you your expertise in what you know and your experience of tile hits, but you are not a structural dynamicist. You are not a stress expert. You are not an image specialist for enhancing images. You are not any of these things. So how can you know?' I said only a larger team of experts could make that kind of determination."

The conversation was going nowhere. Rocha had to leave for the debris assessment team meeting but invited Schomburg to come along and give his perspective to the group. Schomburg agreed.

With only hours remaining before the analysis had to be finished, many of the engineers remained anxious about the lack of a photo. One team member suggested including a slide in their presentation to the MER the next morning restating the critical need for images. Others favored raising the issue not only in the MER, but in the following Mission Management Team meeting. As the discussion continued, however, the idea was dropped.

Schomburg asked Boeing engineers to explain their reasons for wanting the photos. Their reply was straightforward enough: A briefcase-sized object had smacked into Columbia at the relative speed of 500 mph. Schomburg wasn't satisfied. The engineers asked Schomburg why they were bothering with an intensive analysis if shuttle managers already were convinced that the chance of damage was minimal and there was no threat to Columbia. Schomburg responded that some of the engineers were new and the analysis would be a good learning experience.

The team began wrapping up the work late Thursday. With engineers unable to pinpoint exactly where the foam had hit, the resulting 13-page summary considered six possible damage locations. Initially, the study covered four scenarios: the loss of a single tile from three locations on Columbia's belly and minor RCC damage. Two more scenarios covering multiple lost or damaged tiles were added Thursday. All of the potential damage sites were located within a five-sided box that extended from just behind the left wing's leading edge to behind the landing gear door. The tile cases assumed damage down to the dense bottom layer. The lone RCC case looked at possible coating loss to panel 9, located just in front of the projected strike zone.

Some engineers later would say they felt an unspoken pressure to validate management's belief that the foam strike wasn't a flight safety issue. The final presentation explained away some of the more alarming findings. Although the Crater model predicted tile damage would extend down to the dense bottom layer, charts noted that the "program was designed to be conservative due to the large number of unknowns" and often "reports damage for test conditions that show no damage." There was additional safety margin built in elsewhere: Crater analyzed tiles as though they were all the same density and did not account for the fact that the bottom layers were tougher than those at the top. In other words, Crater's predictions were expected to be worse than the damage Columbia had actually suffered.

Nonetheless, one chart noted the "flight condition is significantly outside the test database" and that small changes in the debris' angle of impact and velocity "could substantially increase damage."

Only one of the six cases looked at possible RCC damage. The leading edge panels had a reputation for being tough, and besides, most analysts thought the strike had hit the tiles anyway. Earlier calculations had shown that debris could penetrate the RCC panels if the strike had occurred at a 15-degree angle or greater. The test data used in the formula, however, had been gathered using ice projectiles and not the softer foam. In fact, no study had ever been done to gauge foam damage to RCC. It was simply assumed that foam would be less dangerous than ice. The study predicted that if the RCC panel had been hit, it would have lost only a bit of its outer surface coating. The conclusion: No issue.

Engineers reached the same conclusion for three other scenarios: a missing tile from an access panel adjacent to the leading edge; a missing tile from the main landing gear door; and a missing tile from the lower wing surface. As the team worked into the night, it became clear that two final scenarios involving multiple lost tiles from the landing gear door and lower wing surface would not be completely finished before Friday morning's briefings. The landing gear door case would not be wrapped up until the weekend.

A chart at the end of the presentation summed up the team's findings. While Crater indicated the potential for "large" tile damage, "RCC damage [is] limited to coating" if the debris was foam only. Localized structural damage was possible if a tile was missing, but there would be "no burn through." The study concluded that if the two unfinished cases came back favorably, there would be "safe return indicated even with significant tile damage." Not everyone was comfortable with the team's conclusions, but time was up.

"I still knew," Rocha said, "that we all on the team, including the contractor members, never gave up wanting the imagery of the orbiter."

Mike Dunham, a Boeing engineering manager, e-mailed copies of the presentation that evening to United Space Alliance officials for review. The study was forwarded to Rocha and other NASA officials late Thursday night.

The next morning, when team members briefed shuttle managers on their findings, Columbia's 16-day flight would be half over. Rocha left his office on Thursday with cautious optimism. He was becoming more comfortable with the team's assessment, despite lingering concerns.

The analysis, although still unfinished, so far had indicated that Columbia would return home safely. The conclusions seemed to be conservative. Maybe they were right. After all, the people working on the study were some of the finest engineers in the world.

"I put my faith in the team," Rocha said. "These people were the experts."

S I X

MISSED SIGNALS

We could not believe this was going on, the refusal to support Rocha. I don't know if it was anger as much as disgust. It was like "What in the world is going on?"

—*Johnson engineer Carlisle Campbell*

It was standing room only on the morning of Friday, Jan. 24, as engineers packed the Mission Evaluation Room's conference center to discuss Columbia's foam strike.

Located next to the MER on the ground floor of Johnson's Building 30, the drab, gray room held only 50 people, so the crowd spilled through the doorway and snaked out into the hall. Inside, shuttle managers sat at an L-shaped table in the middle of the room, surrounded by other engineers who occupied chairs that lined the walls on three sides. Engineers at Kennedy and in California were tied in via teleconference.

The meeting began at 7 a.m. sharp. For the next hour, engineers briefed MER manager Don McCormack on two key technical issues: the foam strike and Columbia's heavier-than-normal landing weight. McCormack then headed upstairs to the Action Center to repeat the analysis to the Mission Management Team.

McCormack was a gregarious, straight-talking Texan whom colleagues considered a solid supervisor during his 12 years in the MER. The 48-year-old mechanical engineer started his career in the oil business, then went back to school when the industry went bust in the early 1980s. After graduating from the University of Houston, he landed a job as a thermal analyst with shuttle contractor Rockwell in 1984. He was hired by NASA in 1991 and, for the most part, had worked in the MER ever since. He had been a manager for some 50 shuttle missions.

Fate and scheduling thrust McCormack into the role of lead MER manager for Columbia's flight. The facility was supervised around the clock by a trio of managers working one of three shifts. The two senior MER managers, McCormack and Ken Brown, usually swapped off every mission on the 5 a.m. to 2 p.m. lead shift. The manager on the lead shift ran the pre-MMT meetings and briefed mission managers. It was McCormack's turn.

That morning, the group of engineers included some of Johnson's best and brightest. Rodney Rocha was there, in one of the seats against the wall, as were Paul Shack, Joyce Seriale-Grush, Pam Madera, and McCormack's boss, Fred Ouellette. Calvin Schomburg had come, along with Don Curry, Johnson's resident leading edge expert. United Space Alliance was represented by Loren Shriver and Doug White, director of operations requirements for the orbiter department. Mike Dunham was there on behalf of Boeing and the debris assessment team, along with Carlos Ortiz, who would help make the presentation.

"When I walked into the room that day," McCormack said, "I felt like we had all of the right people doing the work. We had all of the right people in the room to listen to the report."

The foam strike was first on the agenda. The team passed out copies of its analysis. Then Ortiz delivered a background briefing on where the piece of foam came from, its estimated size, and the area on Columbia's left wing where it probably hit. When he finished 10 minutes later, Dunham spoke for another 30 minutes on possible damage to Columbia and what that meant in terms of heating during the ship's re-entry. Many of the managers in the room that morning took comfort from the fact that it was Dunham discussing the issue.

"Mike was not a 'tell you what you want to hear' kind of a guy," one manager in attendance said. "[With] previous analyses he has done, he has brought in bad news—and stuck by it."

Dunham and the team's other engineers still had considerable uncertainty about many of the assessment's conclusions. In particular, they were concerned about Crater's ability to accurately gauge tile damage from a chunk of foam dramatically larger than any used in impact tests. However, that uncertainty never got through to shuttle managers. The charts mentioned that the "program was designed to be conservative due to [the] large number of unknowns" and noted that a "review of test data indicates conservatism for tile penetration." After all, Crater consistently overpredicted damage and didn't factor in the tile's denser bottom layer. Many in the room, including McCormack, interpreted that to mean that there wasn't much to worry about.

"The word 'conservative' was voiced several times," McCormack said, "largely in talking about Crater, saying that Crater was a conservative tool. . . . In hindsight, they almost expected me to be skeptical, but I wasn't. I am thinking to myself, 'Well, hell, I obviously should have been.' But they seemed to be trying to sell me that 'Hey, we think this is a conservative model.' "

Boeing engineers later would say they never intended to leave that impression.

"We certainly weren't selling," Dunham recalled. "We were trying to give a balanced view of what our opinion was and why we thought it was still safe to enter. Uncertainties were part of the briefing. Whether we got the point across or not was difficult for me to ascertain."

One of the two unfinished damage scenarios had been completed late Thursday, and Boeing briefers went over five of the six cases. The remaining scenario, multiple lost tiles on the left wing's landing gear door, would be finished over the weekend. Veteran Boeing engineer Ignacio Norman briefly discussed some of the thermal effects of missing and damaged tile. The team's presentation, based on the skimpy photo evidence, focused almost entirely on the tiles.

Only one chart dealt with a possible RCC strike. Most engineers thought the relative strength of the RCC panels made any impact there a non-issue. In fact, one manager at the meeting was heard to remark that the foam chunk hitting the RCC would be "the best thing that could have happened."

"The general consensus seemed to be that RCC was so tough that this foam couldn't possibly hurt it," McCormack recalled, "and that is kind of what I came away with."

Briefers hurried the presentation to wrap up in time for the 8 a.m. MMT meeting. The bottom line conclusion was on the final page of the charts: "Safe return indicated even with significant tile damage." There was little discussion—and by all indications, little concern.

Rocha sat quietly in his seat. He said nothing.

"One of our thermal engineers who didn't work this problem later told me, 'I never got any sense of concern from anybody. It seemed like you guys were just 'Here is the analysis and you all looked okay to me.' " Rocha recalled. "So, I am being quiet and listening. . . . There are questions going on and 8 o'clock is approaching and there is this pressure: 'Let's wrap it up. We have got to go to Linda Ham's meeting at 8, and we just have got 5 or 10 minutes. Let's speed it up here.' "

As McCormack adjourned the meeting and prepared to brief the MMT, he knew nothing about the debris assessment team's uncertainty or its ongoing anxiety about not getting images of Columbia in orbit. He expressed anger weeks later that no one spoke out.

"That is bull," McCormack said. "Anyone can step up, and if they have a concern, I feel like it can be said without any fear of reprisal. . . . If we have an environment around here where people are afraid to talk, especially a division chief engineer whose responsibility it is to make sure that we program types are all aware of it [an issue], then we have a problem."

Ham convened the third Mission Management Team meeting at 8 a.m. Many of those who listened to the MER presentation, including the Boeing briefers, made the trip upstairs a floor to the Action Center.

Other NASA managers, including Ron Dittemore, participated by teleconference.

The meeting began with a briefing from flight director Phil Engelauf. After a quick discussion about air conditioner problems in Columbia's onboard laboratory, he told the meeting that flight controllers had informed the astronauts of the debris strike. The reason: Reporters had found out.

"We sent up to the crew about a 16-second video clip of the strike just so they are armed if they get any questions in the press conferences or that sort of thing," Engelauf notified other mission managers. "We made it very clear to them, no concerns."

"When is the press conference?" Ham asked.

"Later today," Engelauf replied.

"They may get asked because the press is aware of it," Ham said.

"The press is aware of it," Engelauf agreed. "I know folks have asked me because the press corps at the Cape has been asking."

The meeting moved on. There was a lengthy discussion of Columbia's slightly heavier-than-normal landing weight. Then, a half hour later, the conversation returned to the foam impact. McCormack considered showing the detailed charts he had seen an hour earlier and letting the engineers do the briefing themselves. But there was no requirement for charts at MMT meetings, and that sort of presentation typically wasn't done.

"The thought occurred to me that, hey, I had all these guys and they had their charts. Maybe we should let Linda hear this complete story," McCormack recalled. "But I'm convinced that if she had heard that story, she probably would have gotten to the same place."

McCormack pressed ahead with what would be the Mission Management Team's most extensive discussion of Columbia's foam strike.

"We received the data from the systems integration guys of the potential ranges of sizes and impact angles and where it might have hit, and the guys have gone off and done an analysis," McCormack told the meeting. "They've used a tool they refer to as Crater, which is their official evaluation tool to determine the potential size of the damage. . . . They've done thermal analysis of the areas of where there may be damaged tiles. The analysis is not complete. There is one case yet they wish to run. . . . There's potential for significant tile damage here, but they do not indicate that there is a potential for burn-through. There could be localized heating damage. Obviously, there is a lot of uncertainty in all this in terms of the size of the debris and where it hit and angle of incidence."

Immediately, Ham's thoughts turned to possible delays in preparing Columbia for its next flight. Extensive tile damage meant more processing time would be needed in the ship's hangar to make repairs.

"It's just occurred to me," Ham interjected, "no catastrophic damage and localized heating damage would mean a tile replacement."

"It would mean possible impact to turnaround repairs and that sort of thing," McCormack answered, "but we do not see any kind of safety of flight issue here yet in anything that we've looked at."

"No safety of flight. No issue for this mission. Nothing that we're going to do different. There may be a turnaround [delay]," Ham summarized.

"Right. Right," McCormack replied. "It could potentially hit the RCC, and we don't indicate anything other than possible coating damage or something. We don't see any issue that it hit the RCC. So, although we could . . . have some significant tile damage, we don't see a safety of flight issue."

"What do you mean by that?" Ham asked.

McCormack explained that Columbia could have lost an entire tile, exposing to 2,300-degree temperatures part of a felt pad atop the wing's aluminum airframe that the tiles are bonded to.

"Perhaps," McCormack said, "it could be a significant piece missing but—"

"Would be a turnaround issue," Ham said, completing his sentence.

At that point, Schomburg entered the discussion from the rear of the room. Those on the teleconference couldn't hear as he repeated his oft-stated belief that any tile damage would not be serious—a belief he still held months later.

"Her [Ham's] question was 'Are you comfortable with the analysis that was done on the tile that would say that we had no safety of flight issue on the tile [during] re-entry?' " Schomburg said. "And my answer was yes, and I am even today . . . I would still say that I would be comfortable to fly back if it would hit the tile."

Beyond McCormack's brief remarks discounting the likelihood of RCC damage, there was no discussion of potential problems with the leading edge panels.

"[In] the MER, when they [the assessment team] came to report, they said 'coating damage only,' " Ham reflected later. "There was one sentence on the whole thing [the RCC chart] that they commented on. And the rest went on about the concerns with the tiles."

When Schomburg finished speaking, Ham summarized his remarks for those who couldn't hear over the teleconference connection and wrapped up the discussion:

"We were just reiterating—it was Calvin—that he does not believe that there [will be] any burn-throughs, so no safety of flight kind of issue. It's more of a turnaround issue similar to what we have had on other flights. That's it. All right, any more questions on that?"

Ham scanned the room. It was filled with senior NASA officials, shuttle managers from United Space Alliance, astronauts, engineering managers from the shuttle contractors and NASA, shuttle operations officials and mission scientists. Dozens of others were on the teleconference call. No one spoke up.

Shuttle managers and engineers at Kennedy were listening in a conference room over a speakerphone. Bob Page turned to Wayne Hale and said in disbelief, "That's ridiculous. Is that all they're going to talk about it?"

It was. The short exchange between Ham and McCormack effectively put to rest the issue of Columbia's foam strike for the Mission Management Team. In the team's collective wisdom, it was a tile maintenance problem—nothing more. The meeting ended a few minutes later.

Rocha sat by quietly as he had an hour earlier. He still worried about the lack of photos and the uncertainty in the damage assessment. But now he had a new concern. The assessment team still had not finished analyzing one of the six damage scenarios. Rocha wasn't convinced Ham understood that.

"Our Boeing engineers still had to look at the main landing gear door and the seals around there," Rocha said later. "That was in work. It took two or three more days to finish, but her remark about 'Then I have no safety of flight issue, it sounds like a refurbishment issue like

Calvin said'—the casualness of the remark . . . led me to believe that she didn't get that we still had more work to do."

As the crowd began to file out of the room, Rocha debated going over and saying something to Ham. Astronauts Bob Cabana and Jerry Ross already had stopped Ham with their own concerns about the so-called zipper effect, where the loss of one tile starts a chain reaction of other tiles coming off. Rocha asked Dunham for advice.

"I felt like going in there and interrupting or waiting until they got through and just saying 'Mrs. Ham, I just want you to know that we are not finished,' but I didn't. I didn't do any such thing, so I am waiting," Rocha remembered. "I said, 'Mike, did you hear that she got that we are still not finished?' . . . Mike said to me, 'Well, what are the rules for engaging a manager here? What is the protocol for doing that?' . . . and I remember saying 'Mike, for an issue like this, where we have a flight safety concern, I don't think the protocol should matter. It shouldn't matter at all. We should do it.' But, again, I don't know. . . . He said, 'Oh, I think we told her that there was remaining work,' and I said, 'Okay. I don't know. Maybe she heard it.' I don't know. I don't think she did, but I was doubting myself."

Rocha left without speaking to Ham. Later, as he was leaving Building 30 to go back to his office, he saw Ham getting into her car. Again, Rocha debated approaching her. He kept walking.

Work on the last of the team's six damage scenarios—missing tiles on the main landing gear door—continued during the weekend.

Dunham received the thermal analysis on the effects of the missing tiles and reviewed it on Saturday, Jan. 25. As with the other five cases, the results indicated there would be no flight safety issue. He went over the analysis with debris team chair Madera, who forwarded the results on to Rocha.

Rocha called Boeing engineer Norman on Sunday and reached him on his cell phone in a grocery store. The two went over the final case. Later that night, Rocha called Dunham with a few last questions. Concern was growing among Rocha and other engineers that the final case was potentially the most serious. Hot gas intrusion through a damaged seal on Columbia's landing gear door could destroy the tires and spell disaster during landing. Rocha wanted reassurance from Dunham.

"Tell me why the seal won't disrupt," Rocha recalled asking. "I didn't want to have gas intrusion into the wheel well. That's where the explosive tires are. He went through a rationale I found acceptable."

Rocha signed off on the contractors' findings. At 7:45 on Sunday night, he sent an e-mail to Shack, McCormack, Schomburg, Seriale-Grush and other engineering managers notifying them that the final analysis had been completed:

> Though degradation of the TPS [thermal protection system] and door structure is likely (if the impact occurred here), there is no safety of flight (entry, descent, landing) issue. On Friday I believe the MER was thoroughly briefed and it was clear that open work remained. The message of open work was not clearly given, in my opinion, to Linda Ham at the MMT. I believe we left her the impression that engineering assessments and cases were all finished and we could state with finality no safety of flight issues or questions remaining. This very serious case could not be ruled out and it was a very good thing we carried it through to a finish.

McCormack e-mailed Rocha early Monday to say he would bring the issue to Ham's attention. McCormack mentioned the foam strike for the final time at Monday's Mission Management Team meeting, but only to inform Ham that the last of the six cases had been wrapped up.

"Our results were similar to what we got elsewhere," McCormack told the meeting, "and that is, although local degradation of the door structure is likely if we were to have a sustained hit there, there is no predicted burn-through and no safety of flight issue."

"A non-issue?" Ham asked.

"Yeah, possibly," McCormack replied.

"If it were to hit there . . ." Ham said.

"If it were to hit there," McCormack answered, "it's a critical area on the door, but also the integration guys had indicated that they thought it was a low probability location, but it was still one that we went off and looked at."

"Okay," Ham said.

"So that completes the thermal analysis from the debris hit," McCormack said, "and that's all I've got."

With five days left before Columbia's landing, McCormack's update was the last official discussion of the debris issue by shuttle managers at Johnson. The next morning, Tuesday's daily MER report reflected the fact that the issue was closed:

> The impact analysis indicates the potential for a large damage area to the tile. Damage to the RCC should be limited to coating only and have no mission impact. These thermal analyses indicate possible localized structural damage but no burn-through, and no safety of flight issue.

Shortly after Monday's MMT meeting, Doug Drewry called Rocha and several other NASA engineers into his office. He asked once more if everyone agreed with the analysis and its finding that there was no safety issue. Privately, engineers still wanted a photo of Columbia in orbit.

"Do we have an issue?" Rocha remembered Drewry asking. "We quietly said no. But it's not like zippity-do-da we are all happy now. We are not. We are still in disbelief about what happened, about the photo request. I said, 'Yeah, we did the best we could. We are not identifying a problem.' He said, 'Okay, I just need to know you are all buying into this.' "

After that meeting, Schomburg went to Dittemore to let him know some engineers still wanted a photo. Schomburg felt as certain as ever that the foam could not have seriously damaged the tiles. But he briefly put aside his own skepticism to make Dittemore aware that the clamor for photos had not died.

Dittemore asked why engineers still wanted the pictures.

"I said, 'Well, the bottom line is the uncertainty of where that foam hit,' " Schomburg replied. "His answer, to the best of my recollection, was that 'We have looked at what we can do. We have looked at what [re-entry options] we can fly. We looked at how we can handle this vehicle differently. . . . We don't have a [tile] repair material. We don't have the ability to get out and look at it. We don't have a [shuttle] robot arm. We don't have [spacewalking] tethers. Are you comfortable if it hits tile we are all right?' My answer was 'Yes, I am still comfortable that we would be all right in the tile world.' He said that he wasn't going to recommend taking the photo."

With the analysis over, chairwoman Madera congratulated the debris assessment team's engineers.

"Wanted to forward a note of thanks for a job well done—a direct reflection of your hard work and significant contributions!" her e-mail said. "Please accept my thanks as well—an excellent job by all!"

Congratulations notwithstanding, some engineers still were worried. On Monday afternoon, discussions continued about the landing gear door. John Kowal, a NASA thermal engineer, sent out an e-mail clarifying Norman's analysis on the sixth and final scenario:

> In the case he ran, the large gouge is in the acreage of the door. If the gouge were to occur in a location where it passes over the thermal barrier on the perimeter of the door, the statement that there is "no breaching of the thermal and gas seals" would not be valid. I think this point should be clarified; otherwise, the note sent out this morning gives a false sense of security.

The thermal seal was a crucial component. The barrier, made of silicon with a flame-resistant Nomex coating, kept super-hot gas from working its way into the wheel well and damaging the landing gear or its tires.

Carlisle Campbell, a 73-year-old NASA structural engineer at Johnson, was among the recipients of Kowal's e-mail. Campbell left an air-conditioning company in his native North Carolina to take a job with NASA in 1962. After helping design parachute systems for the Gemini and Apollo capsules, he began working on the shuttle's landing gear in 1974. He had become an authority on the system in the three decades since.

The growing fear among Campbell and other engineers wasn't that Columbia would suffer a massive burn-through that would destroy the ship, but that enough heat would get into the wheel well to damage the landing gear or flatten its tires.

"The reason for the two-flat-tire scenario is that there are some temperature-sensitive safety devices in the shuttle wheels, as well as all airliner wheels," Campbell said. "They're called thermal fuse plugs. They let go [let the air out] at pretty low temperatures. Ours let go at 340 degrees Fahrenheit. That's not very hot. I was afraid that if we got a plasma leak and entry heating, these plugs might release the tire pressures. So we would be landing on two flat tires."

Campbell had read Rocha's Sunday e-mail outlining the team's sixth and final damage scenario. The conclusions left Campbell more concerned than ever.

"They said we might have a little bit of a plasma leak into the wing and it won't hurt much as long as it doesn't get to a sensitive area like the landing gear area or the wheel well," Campbell said. "To me that was a pretty dangerous scenario. Yet these thermal guys had that in their presentation to our management. Management saw those words, and they didn't take it the way I did. I can't imagine why they wouldn't take that as a devastating warning. . . . We got angry. What were these guys thinking?"

Campbell also had been angered by daily updates he was receiving from Rocha on management's indifference to the photo request. He forwarded Rocha and Kowal's e-mails, along with briefing slides and images of the debris strike, to Bob Daugherty, an old friend at NASA's Langley Research Center in Hampton, Virginia. Daugherty was another expert on the shuttle's landing gear. The two had known each other for almost 20 years.

Daugherty had heard nothing about the foam strike and was amazed to read the messages. After watching video of the impact, he e-mailed Campbell on Monday afternoon about whether it was best to have Columbia's astronauts bail out or attempt a belly landing in the event the landing gear or tires were damaged.

"WOW!!! I bet there are a few pucker strings pulled tight around there! Thinking about a belly landing versus bailout. . . . I would say that if there is a question about main gear wheel burn thru that it's crazy to even hit the deploy gear button," Daugherty wrote.

Shuttle pilots lower the landing gear at an altitude of 300 feet, about a half minute before touchdown. One of his worries was that damage to the tires and the aluminum wheels they were mounted on might prevent the left landing gear from deploying properly: "300 feet is the wrong altitude to find out you have one gear down and the other not down. . . . You're dead in that case."

Daugherty added that the prospect of a wheels-up runway slap-down seemed so scary that "I would bail out before I would let a loved one land like that."

Campbell e-mailed back a few minutes later: "That's why they need to get all the facts in early on—such as look at impact damage from the spy telescope. Even then, we may not know the real effect of the damage."

No one seemed to know for sure whether the crew could survive a landing attempt with two flat tires. The orbiters put up to 140,000 pounds of stress on the tires while landing at speeds up to 225 mph. That's twice the stress on the tires of a 747 jumbo jet. The concern was that tire problems could cause Husband to lose control of Columbia on the runway. Campbell and Daugherty decided to find out.

As luck would have it, shuttle landing simulations were going on that week in a specially equipped facility at Ames Research Center. The pair contacted Howard Law, a Johnson engineer who was at Ames helping run the simulations.

"One of the things they were looking at was . . . if you have one tire flat . . . could the other tire on a strut survive if its companion is at some reduced pressure or flat?" Daugherty said. "So, we were running tests that very week. . . . Strangely enough, they were actually exercising part of the software that you would want to use if you wanted to look at two flat tires. Quite a coincidence."

There were, however, problems. Astronauts were using the simulator for training runs during the day. And, like the ill-fated photo request, there was no official directive from shuttle managers in Houston.

"They did not want to do the work right then because they didn't have permission from the project manager," Campbell recalled. "We said, 'Hey, we don't have time to go through the red tape to get permission. We've just got to do this because it's necessary in our opinion.' They said, 'OK, we'll try to work it into the schedule and it won't cost any more.' "

By Tuesday, Daugherty and Campbell were calling and e-mailing each other several times daily. Daugherty asked Campbell in a message if there was "Any more activity today on the tile damage or are people just relegated to crossing their fingers and hoping for the best?" Daugherty also sent a message to his boss, Mark Shuart, director of Langley's structures and materials branch, after Shuart caught wind of what was going on. But Daugherty's e-mail to Shuart downplayed the seriousness of the situation:

"So far, our involvement has been one of providing the current model of drag associated with landing with two tires flat prior to touchdown and some thought exercises of what might happen if the wheel well were burned into . . . something that is arguably very unlikely." Because other simulations with one flight tire were going on, Daugherty concluded, "It is a very convenient time to look at two tires flat if they can squeeze it in. Will keep you informed as I hear more . . . if I do."

Daugherty, Campbell and Law spoke by conference call Wednesday morning. The two landing gear experts stressed flight controllers would need the information if Columbia's tires had problems. Mission Control would be ready to instantly relay instructions to the crew during a possible emergency. The engineers decided they would try to stage up to a half-dozen simulations after hours at night.

That afternoon, Daugherty e-mailed Shuart again, this time, with his frustration and an update on their progress:

> It is as simple as hitting a software button and simply doing it. But since no Orbiter Program Management is "directing" the sim[ulation] community to do this, it might need to get done "at night." An anecdote they told us is that this was already done by mistake this week and the commander lost control of the vehicle. . . . It seems that if Mission Operations were to see both tire pressure indicators go to zero during entry, they would sure as hell want to know whether they should land

> gear up, try to deploy the gear, or go bailout. We can't imagine why getting information is being treated like the plague.

By now, the issue was moving up the chain of command at Langley. On Thursday, Shuart forwarded Daugherty's e-mail to Doug Dwoyer and Ruth Martin, two NASA executives in the Langley director's office. News of the development got back to Johnson's engineering managers that same day. Shack sent a "high importance" message to Frank Benz, Seriale-Grush and other engineering managers about Daugherty and Campbell's activity. The e-mail said little about potential landing issues but expressed considerable concern about the possible involvement of NASA Headquarters in Washington.

"Just a heads up on possible high level management activity—The Langley Center Director heard about the debris impact and the action's being worked," Shack's e-mail said. "He is writing to Bill Readdy to offer Langley's help since they are landing gear experts, in spite of there being no technical concern based on the analyses that have been completed."

Shack's e-mail included an excerpt from Rocha's Sunday message describing how the analysis had predicted no burn-through of the landing gear door, no breaching of the thermal seals, and no warping or deformation of the door that could let hot gases blow inside. The message repeated Rocha's conclusion there was no safety of flight issue. "What goes up must come down, so I did give Ralph [Roe] a heads up that he may hear from HQ."

Meanwhile, Campbell and Daugherty had decided to contact flight controllers in Johnson's Mechanical Systems Group to let them know what was going on. Daugherty drafted a lengthy note on Thursday afternoon covering virtually every contingency he could think of. He e-mailed it to David Lechner, the flight control team's mechanical systems lead engineer for landing hardware:

> I'm writing this e-mail not really in an official capacity, but since we've worked together so many times I feel like I can say pretty much anything to you. And before I begin I would like to offer that I am admittedly erring way on the side of absolute worse case scenarios and I don't really believe things are as

> bad as I'm getting ready to make them out. But I certainly believe that to not be ready for a gut-wrenching decision after seeing instrumentation in the wheel well not be there after entry is irresponsible.

The message detailed seven issues Daugherty thought the landing team should be prepared to deal with: What if the tires exploded and blew the landing gear door open? What if heating prematurely fired explosive devices used to open the doors? What if hot gases damaged hardware needed to lower the landing gear? What happens if one landing gear deploys and the other doesn't? Can you land Columbia on its belly? What about having the crew bail out? Could Husband control Columbia on the runway with two flat tires?

Early Friday morning, Lechner was at Mission Control for a standard checkout of Columbia's systems that is always done the day before landing. He forwarded Daugherty's message to other members of Johnson's flight control team. The e-mail generated a spirited discussion. One engineer, Bill Anderson, asked, "Why are we talking about this on the day before landing and not the day after launch?" Another, Kevin McCluney, gave a hypothetical description of what data might appear on flight controllers' computer consoles if the wheel well was breached. This scenario later would prove prophetic—although not in the way McCluney had intended.

> Let's surmise what sort of signature we'd see if a limited stream of plasma did get into the wheel well (roughly from EI [entry interface] + a few minutes until about 200,000 ft.; in other words, a 10 to 15 minute long window). First would be a temperature rise for the tires, brakes, strut actuator, and the uplock actuator return. Tire pressure (and theoretically brake pressures as the fluid temperature increased, though the expansion is small) would rise given enough time, and assuming the tire(s) don't get holed. Then the data would start dropping out as the electrical wiring is severed. . . . Data loss would include that for tire pressures and temperatures, brake pressures and temperatures.

McCluney's message pondered what the alert flight controller should do if he or she saw those indications. He wrote:

> There are only a limited number of choices. (1) Do nothing. Assume it's just a bunch of smart transducer failures or that the gear can take the punishment. (2) Decide that the gear is probably toast, call for an early enough deploy to allow for a bail-out if required [less than Mach 0.9] and rely on the (remaining) data or video in order to decide between a bail-out and a landing attempt. (3) Decide that the gear is toast, that landing is impossible, and call for a bail-out. (4) Decide that the gear is toast and recommend a gear-up landing.

Option 4 probably would mean losing the crew and vehicle, McCluney noted. Option 3 seemed extreme because it meant throwing away Columbia without lots of supporting data. The choices, he concluded, were between options 1 and 2. How many measurements disappeared would be the determining factor.

Campbell e-mailed results from the Ames simulations to Daugherty and several Johnson engineers on Friday that supported McCluney's option 1. Two pilots had flown six landings. Although the runs were done on short notice and had some inaccuracies, the results were thought to be "generally plausible."

"The results showed that this condition [two flat tires] was survivable/controllable," Campbell wrote. "Even nose slapdown was within limits."

Daugherty and Campbell reviewed the results late Friday afternoon in a teleconference with Bob Doremus and David Paternostro, two managers in mission operations at Johnson. Doremus and Paternostro remained skeptical that landing with two flat tires was survivable. Nevertheless, they thanked Campbell and Daugherty for the information. If the worst case did materialize, flight controllers would have a better handle on their options.

Campbell went so far as to call the control tower at Kennedy's shuttle landing strip to ask managers if they had the ability to foam the runway during an emergency. They didn't. Nevertheless, no one, including Campbell and Daugherty, actually expected the worst to happen as they left work on Friday night just hours before Columbia's landing.

"I thought it was a real, but low, probability," Campbell said. "I thought we would have, at worst, what the thermal guys were predicting—a small burn-through. We might have damage to some of the landing gear, or we might not. I slept all right. I wasn't worried about a massive catastrophe."

As Columbia's second week in space wound down, concern about the debris strike also had reached NASA Headquarters in Washington.

On Jan. 22, Mike Card, a manager in the headquarters' Safety and Mission Assurance Office, stopped by to tell Bryan O'Connor, NASA's top safety chief, about an earlier discussion with an acquaintance. Someone had contacted the National Imagery and Mapping Agency to look into getting photos of Columbia in orbit after the foam strike during launch. Card spoke with Johnson's safety office, which informed him the event was "in family." But a friend with the intelligence agency wanted Card to know the pictures could be taken if NASA wanted them.

O'Connor already knew about the impact. A copy of the Johnson film team's report had been forwarded to him the day before.

"[Card] was just asking whether this was a need or not because Mike isn't in the shuttle chain of command," O'Connor remembered. "So I read his question as 'Was I aware of any requests that were being made?' My answer to him was that the [shuttle] program will exercise this relationship at the appropriate point when they feel like they have a need. You don't have to be going through your friendship or some other loop because they have a [formal process]."

A week later, on the Wednesday before landing, Card broached the issue again, this time with Readdy, NASA's top official for manned spaceflight. The military was willing to take pictures of Columbia. All NASA had to do was ask. Readdy recalled that as the two continued talking, it became obvious that Card wasn't aware an analysis had concluded there was no safety concern. Readdy retrieved a copy of the MER report from the previous day showing the issue had been put to rest.

"[Card] read it and he said, 'Well, you know, they are still saying that they would be willing to go do this. All you have to do is change the priority. If you declare it a spacecraft emergency, then they can go off and go turn on the request,' " Readdy remembered Card saying. "At that point, I told him 'Routine is the best we can do here. Right here in this report, it tells me there is no safety of flight issue, and the team, I know, has worked this very diligently over the course of a long period of time. They have not arrived at this conclusion arbitrarily.' "

Card left. It was the last discussion of the photo request among NASA managers in Washington. Campbell, however, tried one last time to bring concern about the foam strike to the attention of senior agency officials two days before landing.

After considerable deliberation, Campbell took the bold step of forwarding to O'Connor a copy of Rocha's e-mail from the previous Saturday. Campbell had known O'Connor since the safety chief's astronaut days, when the two had worked together on shuttle tire issues after the Challenger accident.

"I asked Rocha several times during that week to contact O'Connor," Campbell recalled. "He finally said, 'I'm just not going to do it. That's jumping the chain of command. That's too risky.' So I did it. Of course, I didn't tell him I did it. I did confide in Daugherty and a couple of my previous supervisors. They all said I did the right thing."

Campbell's brief note said, "This is confidential but I just wanted to be sure that you were aware of the potential landing gear door damage." Rocha's attached message detailed possible issues with the landing gear door analysis. But it also concluded there was no safety of flight issue and expressed relief that the analysis had been completed as planned.

O'Connor was already familiar with the outcome of the analysis. He saw little to be alarmed about in Campbell's note. He was surprised, he replied, at the degree of analysis required, thanked his old friend for the message, and asked how he was doing.

"I look at that now in retrospect," O'Connor said, "and wish I had just called him and said, 'What is "confidential" about that? Are you trying to say something to me more than what I see here?' Maybe he was. I don't know."

In fact, Campbell was gambling that his e-mail would spur O'Connor into some sort of action.

"What I had hoped—and it didn't happen—was that he would read Rocha's memo and say 'Good grief. What are these guys saying down there? What is going on?' " Campbell remembered. "I thought he would jump on it. I had hoped he would investigate it further by either calling Rocha or something to that effect. . . . But I was a poor messenger. I didn't have any proof."

Campbell replied to O'Connor's brief note on Friday. "We are doing great, but it's not like the good old days. The LaRC [Langley] Center Director plans to offer their help on the debris impact concern through Bill Readdy, I have heard. We don't have much time left if there is another issue."

The e-mail exchange ended there.

Like Campbell, Rocha still had concerns that Friday evening but didn't expect the worst the next morning. Even so, as a precaution, he asked MER engineers to set up a computer display where he could monitor temperature and tire pressure readings during re-entry from Columbia's left wing and landing gear, and compare them with readings from the right wing.

"I just wanted to look for anomalous trends or anything interesting between left and right," Rocha said. "When I got in the next morning, on Saturday, they were ready."

Rocha's concerns remained focused on the tiles around the landing gear door. No one was worried about the leading edge or its reinforced carbon-carbon panels.

At Kennedy, landing crews prepared to arrive early the next morning for another routine landing. Atlantis had been moved into the Vehicle Assembly Building to be attached to its tank and boosters for roll out to the launchpad the following week. Shuttle managers had debated several days earlier about whether to delay the move. Major foam loss from the bipod ramp now had been seen during two of the three previous launches.

"We argued about it," Schomburg said. "The answer was that they were going to roll it out." The move took place, he noted, without "doing anything different to the ramp."

Some 184 miles above Earth, Columbia's crew remained oblivious to the drama that had unfolded below during the previous two weeks. The seven astronauts were aware of the foam strike because of the brief note from flight controllers the week before. But it was nothing to worry about, they had been assured. They knew nothing of photo requests, debris assessments or detailed discussions of blown tires and possible landing disasters.

Columbia's astronauts had no worries. Tomorrow, they were coming home.

TOP: *Dec. 9, 2002: An unidentified observer in the Launch Control Center at the Kennedy Space Center looks on as shuttle Columbia is hauled from the Vehicle Assembly Building to pad 39A for a January launch. It would be the 28th flight for NASA's oldest space shuttle and the 113th since Columbia's maiden voyage in April 1981.*

LEFT: *An accident waiting to happen. The piece of external fuel tank insulation that ultimately doomed Columbia and its crew broke away from an aerodynamically shaped ramp of insulating foam at the base of a bipod strut anchoring the shuttle's nose to the ship's external fuel tank (upper circle). The foam hit the leading edge of the shuttle's left wing at reinforced carbon-carbon panel 8 (lower circle).*

ABOVE: *Jan. 16, 2003: The Columbia astronauts pose for a traditional launch-day "photo op" at crew quarters at Kennedy. From left to right: Israeli payload specialist Ilan Ramon; physician-astronaut Dave Brown; pilot William "Willie" McCool; mission commander Rick Husband; physician-astronaut Laurel Clark; payload commander Mike Anderson; flight engineer Kalpana Chawla. Husband, Anderson and Chawla each had one previous flight to their credit. The rest were rookies.*

RIGHT: *The astronauts depart crew quarters and head for Launch Complex 39A. Shuttle Columbia was fueled and ready for take off on a 16-day science mission featuring more than 80 experiments.*

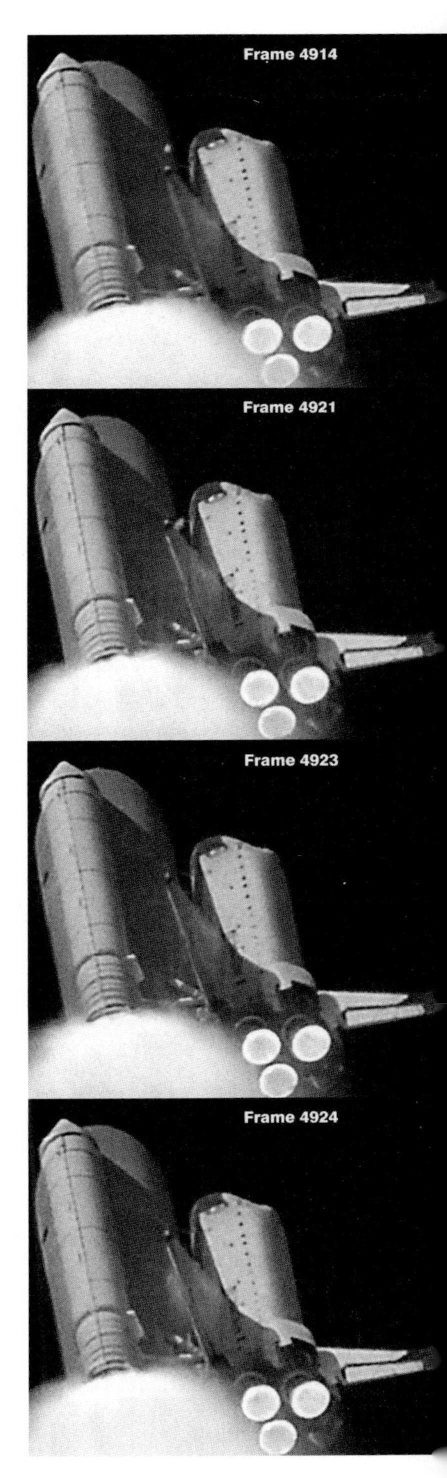

ABOVE: *10:39 a.m., Jan. 16, 2003: Right on time, shuttle Columbia roars away from its oceanside launchpad. Note the wedge-shaped left-side bipod ramp under the shuttle's nose. A little more than a minute later, a large piece of the ramp would break away and hit the shuttle's left wing.*

RIGHT: *10:40:22 a.m., Jan. 16, 2003: A briefcase-sized chunk of foam insulation breaks away from the left bipod ramp of Columbia's external fuel tank 81.7 seconds after liftoff as seen in these enhanced video frames from a NASA tracking camera. The shuttle's velocity is 1,568 mph, and the foam breaks into several pieces as it tumbles in the air. In two-tenths of a second, the largest piece of debris slows to 1,022 mph as it disappears behind Columbia's left wing (Frame 4923). It emerges in a powdery looking shower of debris after hitting the wing at a relative velocity of about 545 mph.*

Engineers would mistakenly conclude the 1½-pound piece of foam hit the underside of the wing at a glancing angle and that whatever damage might have resulted was not a threat to the orbiter or its crew. In reality, the foam slammed into the lower side of the wing's leading edge, blasting a 6- to 10-inch-wide hole in reinforced carbon-carbon panel 8. As investigators would later show, the hole provided a path for super-heated air to enter the wing during re-entry.

Columbia's crew was not immediately aware of the foam strike or its possible consequences. The goal of the mission was 16 days of research involving scores of high-tech experiments, many of them located inside the relatively roomy Spacehab research module mounted in the shuttle's cargo bay. Other experiments were mounted outside, behind the Spacehab module, in the vacuum of space. In this fisheye view, Chawla looks back through the tunnel connecting the module to the shuttle's crew cabin.

The foam debris that fell from Columbia's fuel tank is believed to have struck leading edge panel 8. The panels are numbered from 1—closest to the fuselage—to 22 at the tip of the wing. As this frame from orbit shows, the astronauts could not see panel 8 from their cabin

TOP TO BOTTOM:
Frames from a videotape shot during Columbia's descent Feb. 1 show Husband, McCool, Clark and Chawla working through their re-entry checklist as the shuttle falls into the discernible atmosphere 400,000 feet above the Pacific Ocean. In the first frame, McCool adjusts the camera while Husband, to his left, monitors Columbia's instruments. McCool then hands the camera to Clark, sitting directly behind him on the right side of the cockpit, who coaxes Chawla to wave and smile. Clark then turns the camera on herself. At this point, there are no signs of anything amiss and entry is proceeding normally. This heat-damaged videotape, found beside a country road near Palestine, Texas, ended before the crew had any awareness their lives were in danger.

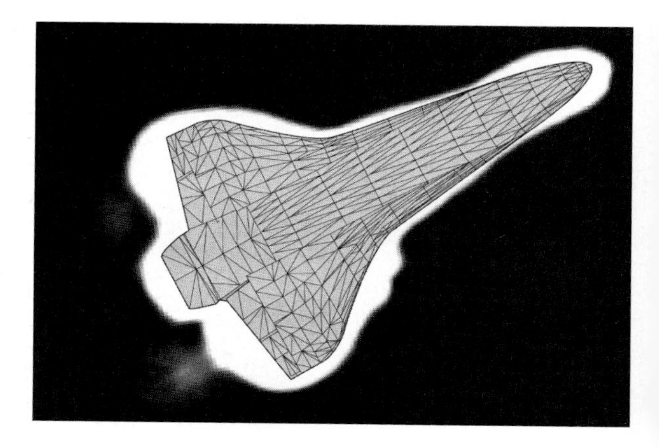

8:57 a.m., Feb. 1, 2003: Off-duty engineers at Kirtland Air Force Base, N.M., using a small 3.5-inch telescope and a computerized tracking system, photograph Columbia as it streaks toward Texas. **LEFT:** *Looking at Columbia from below, the leading edge of the left wing appears deformed and an unusual plume trailed the spacecraft.* **RIGHT:** *The shuttle's actual shape superimposed on the blurred image. This photo was taken roughly three minutes before all or a large portion of the left wing failed and broke away.*

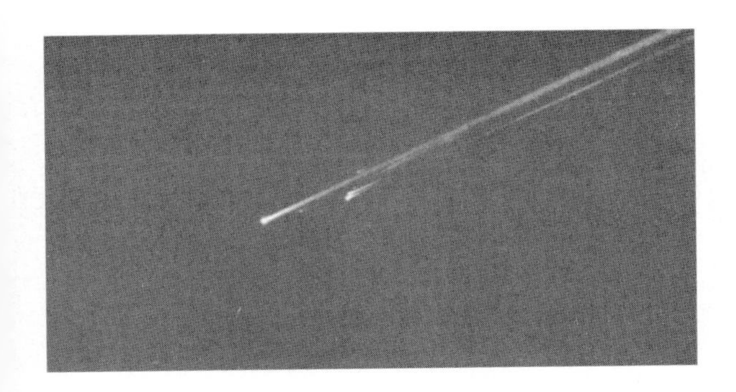

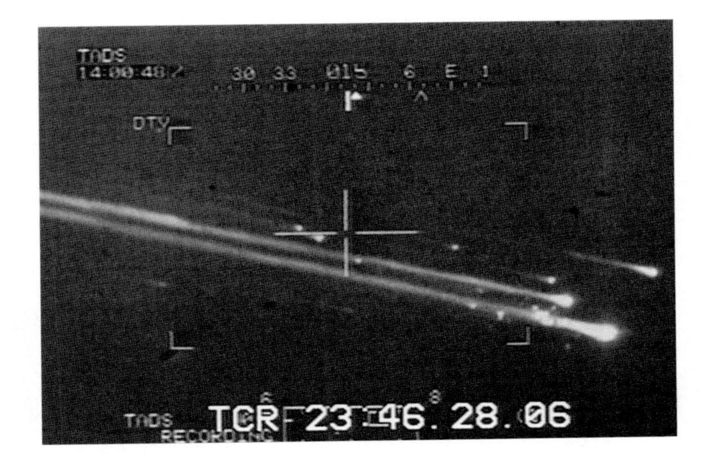

9:00 A.M., *Feb. 1, 2003: Veteran shuttle watchers along Columbia's ground track turned out in force to watch the shuttle's fiery re-entry. Due to the timing of the entry, the shuttle was visible from the coast of California all the way to central Texas.* **LEFT:** *In this view from a Dallas suburb, photographed by space enthusiast Jim Dietz, Columbia has already broken up into multiple pieces of smoking debris. "You could see it tumbling," Dietz said. "And the contrail behind it would get wider, narrower, wider, narrower. Like that. I was shaking like a leaf. I knew what had happened."* **RIGHT:** *The view from an Apache helicopter near Fort Hood, Texas.*

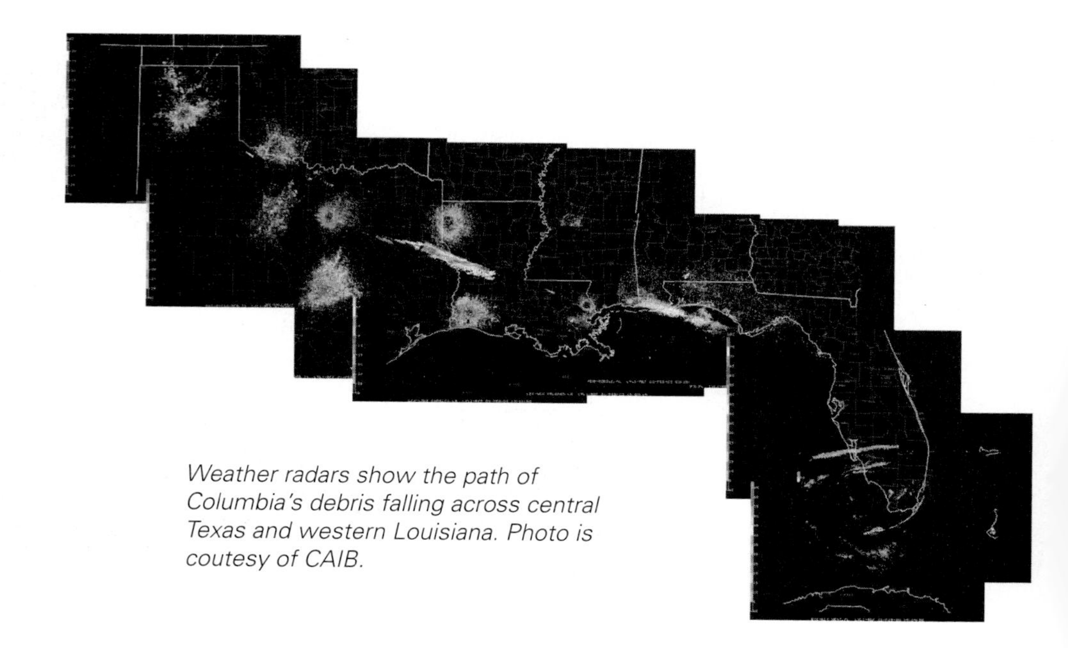

Weather radars show the path of Columbia's debris falling across central Texas and western Louisiana. Photo is coutesy of CAIB.

9:15:04 a.m. EST, Feb. 1, 2003: Mission Control at the Johnson Space Center. On a large tracking map, Columbia's position is frozen above central Texas. An "S" follows each numeric entry on the left-hand screen, meaning the data are "static" and not changing. The shuttle had been scheduled to land at the Kennedy Space Center at 9:16 a.m. and the reality of Columbia's demise is sinking in with brutal finality.

Michael Kostelnik, deputy associate administrator of the shuttle and space station programs, confers with Bryan O'Connor, a former shuttle commander in charge of NASA's safety program, at agency headquarters in Washington. O'Connor was one of the few senior managers to question the "safety of flight" rationale justifying shuttle launchings in the wake of a major foam shedding "event" during launch of the shuttle Atlantis in October 2002.

Former shuttle commander William Readdy, chief of manned spaceflight (left), NASA Administrator Sean O'Keefe (center) and Deputy Administrator Frederick Gregory, also a former shuttle commander (right), follow the aftermath of the Columbia disaster at NASA Headquarters in Washington. O'Keefe and Readdy began their day at the Kennedy Space Center, awaiting Columbia's return to Earth just a few feet away from the families of the astronauts. O'Keefe's first inkling of disaster was the ashen look on Readdy's face after flight controllers lost contact with the shuttle.

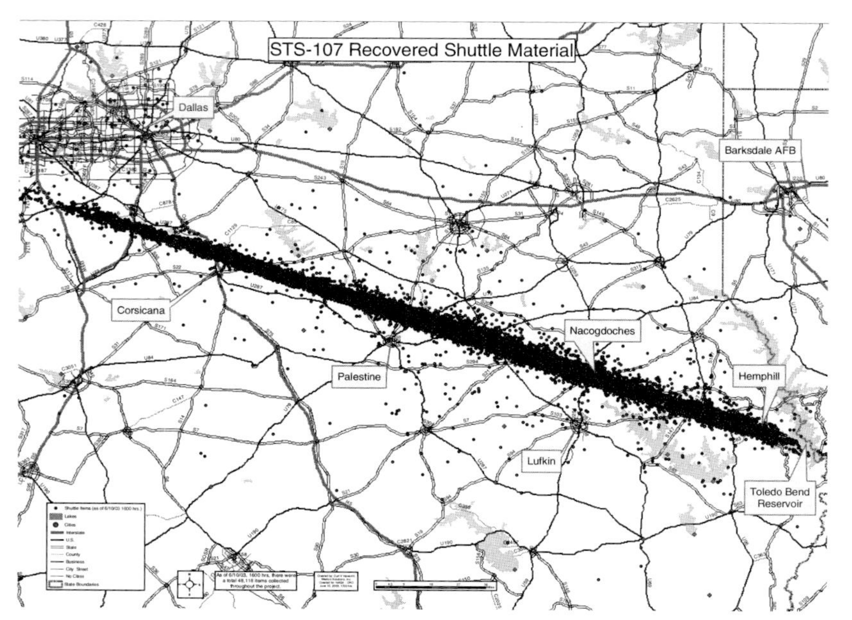

Wreckage from Columbia fell over a 2,000-square-mile "footprint" stretching across central Texas and extending into western Louisiana. By the end of the first day, local emergency agencies were receiving debris reports at the rate of 18 calls per minute and President George Bush declared east Texas a federal disaster area. Remarkably, no injuries were reported.

Volunteers and emergency response personnel search a field in Navarro County, Texas, looking for shuttle debris. NASA and the Federal Emergency Management Agency called in Forest Service fire fighters to assist in a massive "grid search," ultimately recovering more than 1,000 pieces of debris per day. All told, more than 25,000 men and women from 270 organizations and agencies participated in the debris recovery effort, covering more than 2.3 million acres. Some 84,900 pounds of orbiter wreckage were recovered.

Residents of Hemphill, Texas, erected a simple memorial to mark the location where remains of a Columbia astronaut fell to Earth.

In the hours after the accident, residents in the Houston area began leaving flowers, banners and cards at the main entrance to the Johnson Space Center, creating an impromptu memorial. Over the days and weeks that followed, the memorial grew to stretch nearly a block along NASA Road 1.

Feb. 4, 2003: President Bush attends an emotional memorial service at the Johnson Space Center three days after the mishap. "The final days of their own lives were spent looking down upon this Earth. And now, on every continent, in every land they could see, the names of these astronauts are known and remembered. They will always have an honored place in the memory of this country. And today I offer the respect and gratitude of the people of the United States."

The astronauts' remains arrive at Dover Air Force Base, Del., on Feb. 5. The search for human remains was a grim but quickly completed task. By Feb. 13, remains of all seven crew members had been positively identified by Air Force pathologists at Dover.

The Columbia astronaut families, friends and supporters participate in an April 16 tree planting memorial at the Johnson Space Center.

RIGHT: *Columbia comes home. A truckload of shuttle debris, collected in Texas and shipped to Florida from Barksdale Air Force Base in Louisiana, arrives at the Kennedy Space Center for reconstruction and analysis. The wreckage will be processed at an unused hangar next to the 3-mile-long shuttle runway where Columbia's crew had hoped to land.*

BELOW: *Taking advice from the National Transportation Safety Board, NASA investigators set up a grid in the reconstruction hangar.*

ABOVE: *Within a week of the disaster, the grid was in place. By the end of May, nearly 38 percent of the shuttle had been recovered. Only critical wing and fuselage wreckage was actually placed on the grid.*

LEFT: *A walled-off corner of the reconstruction hangar keeps cockpit wreckage and recovered crew equipment from prying eyes. The outer walls are adorned with cards, notes and signatures from school children and well wishers.*

Smart, polished and well spoken, shuttle program manager Ronald Dittemore was a charismatic leader who reassured the public in the days following the mishap. Criticized by some for his management style, others praised his ability to juggle complex technical issues. Dittemore never hesitated to delay launch if there was a safety issue. But even he missed the threat posed by external tank foam shedding.

Linda Ham briefs the newly formed Columbia Accident Investigation Board on Feb. 7 in her short-lived post-accident role as chairman of NASA's Mishap Response Team. Ham served as a lightning rod for criticism leveled against NASA in the weeks and months after the accident, in part because of her uncritical acceptance of a hurried, ultimately incorrect analysis of the foam impact danger.

Bill Readdy, NASA's associate administrator for space flight and the man ultimately responsible for shuttle operations, inspects the recovered tip of Columbia's tail fin in the reconstruction hangar at Kennedy. Readdy offered his resignation to NASA Administrator Sean O'Keefe within hours of the accident. O'Keefe refused.

Retired Admiral Harold Gehman, chairman of the Columbia Accident Investigation Board, brought keen intelligence, organizational skills and political savvy to the disaster probe, winning widespread respect from fellow board members, reporters and Congress. Gehman decided early on not to assign individual responsibility for the human failures that led to Columbia's demise, instead leveling criticism at NASA's safety "culture." Photo is coutesy of CAIB.

ABOVE: *On March 19, a search crew near Hemphill, Texas, made the single most important recovery in the salvage operation: a data recorder loaded with readings from some 721 sensors that operated during launch and re-entry. Remarkably, the recorder was in virtually pristine condition and when engineers opened the device they were elated to discover the tape intact. Based primarily on this recovered data, investigators ultimately were able to develop a detailed failure scenario.*

BELOW: *Columbia's two main landing gear struts and tires in the reconstruction hangar. The left outboard tire, the one subjected to the most extreme temperatures as a jet of super-heated air burned its way into the landing gear wheel well, is in the foreground.*

ABOVE: *Columbia's forward cockpit window frames, a poignant reminder of the forces the shuttle endured in its final moments. Three roses left by astronaut family members are visible on the window frame support pallet.*

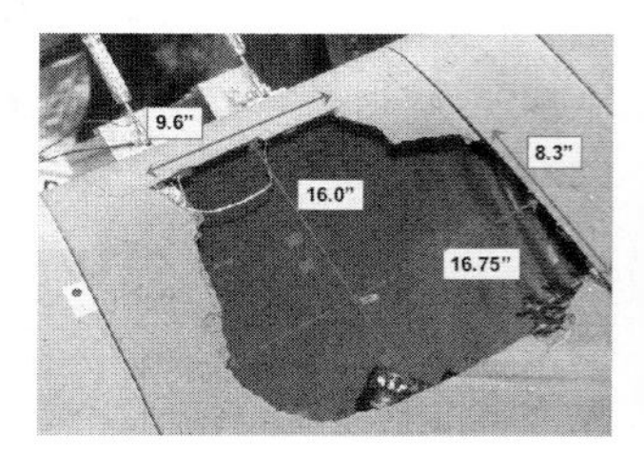

In a critical test, a chunk of external tank foam insulation was fired at a shuttle wing mockup at the Southwest Research Institute in San Antonio, Texas, using a nitrogen gas cannon. This test allowed the Columbia Accident Investigation Board to conclude, without qualification, "the foam did it."

A lingering question to the Columbia accident investigation was what, if anything, NASA could have done to save Husband and his crewmates. The Columbia Accident Investigation Board concluded the crew could have been saved, but only if two assumptions were met: That engineers knew early on, without any doubt, the shuttle had suffered a catastrophic breach and if agency managers would have committed a second shuttle and its crew to a rescue flight without knowing precisely what doomed Columbia. In this computer graphic, the shuttle Atlantis is positioned below Columbia for a spacewalking crew transfer.

SEVEN

DISASTER

We didn't hear anything and we kept waiting and nothing. Then we knew it wasn't going to be good. It was approaching landing time and we got to landing time and there is no orbiter in site. We waited a couple more minutes and said 'You know, this is it. They are not coming to Florida today.'

—Bob Cabana, astronaut and director of flight crew operations at Johnson.

Jay Lawson, a defense contractor and an amateur astronomer in Sparks, Nevada, watched the shuttle approach from his front yard. He had never seen a shuttle entry before and had spent two days getting ready, downloading charts from NASA showing the shuttle's track across the sky, making sure he had the timing right and sending e-mails to friends urging them not to miss the show. Shuttles rarely re-entered over Sparks, and the weather was ideal.

The sky maps showed Columbia would pass almost directly in front of Venus. Lawson had been experimenting with a video camera, shooting the planets and the moon through his telescope, and decided to simply hand-point the camera at the shuttle and follow it across the sky. He didn't know what to expect.

But just like clockwork, Columbia appeared in the west, "incredibly obvious and bright and pretty much hauling ass across the sky," he said. It looked like a slow-moving shooting star, "but a lot more orangish kind of color and obviously, much larger and brighter" than any he'd ever seen before.

The video captured his excitement.

"Wow!" Lawson exclaimed as he tracked Columbia in the camera's flip-out viewfinder. He was using an old astronomer's trick, watching the viewfinder with one eye and the sky with the other. As the shuttle approached Venus, he zoomed in. And suddenly the shuttle, almost as bright as the planet it was approaching, briefly flared in intensity. The flare was unmistakable. And even to his untrained eye, it did not look normal.

"Whoa!" Lawson said to himself, his voice picked up by the camera. "Holy crap, that was weird."

When the shuttle disappeared from view to the east, Lawson ran inside his house and plugged the camera into a television set. A loud sonic boom rattled his windows as he replayed the tape. He again saw the shuttle's sudden flare and then noticed something he'd missed earlier. Just after the flare, what appeared to be a large piece of debris separated from the shuttle and fell behind in the ship's plasma trail. Whatever it was, it remained visible for several seconds, glowing in the dark sky before slowly fading out.

"That's odd," he thought. "What's that coming off the back?"

He switched his TV to the NASA channel and listened in as commentator James Hartsfield described what seemed to be a normal descent toward Florida. He pushed the record button on his VCR.

Eight hundred miles to the southeast, four satellite tracking experts at Kirtland Air Force Base just outside Albuquerque, New Mexico, had mounted a digital camera on a small telescope. The telescope was aimed at a computer-controlled mirror system that was programmed to track the shuttle. The engineers wanted to take a few shots of Columbia as it flew past to test the software. Tracking an extremely fast-moving aircraft with what amounted to a large, zoomed-in telephoto lens was a difficult challenge. But the engineers were experts. They worked at Kirtland's Starfire Optical Range, which specializes in satellite tracking and telescopes designed to counteract the distortion caused by the atmosphere.

As Columbia streaked past, the computer aimed and moved the mirrors, and the camera captured a string of pictures. Columbia's general shape was distinct but blurred by the effects of extreme heating and the shuttle's speed. Even so, the pictures showed what appeared to be turbulence of some sort around the leading edge of the left wing. What looked like a vapor trail extended away behind the wing. No one had ever taken a close-up picture of a shuttle during this phase of re-entry, and the engineers weren't sure whether they were seeing anything unusual or not. But the left wing definitely looked different from the right. They didn't know it then, but their pictures would run in magazines and newspapers across the nation in just a few days.

In Mesquite, Texas, a suburb on the eastern side of Dallas, Jim Dietz was watching Columbia from his backyard. A long-time space enthusiast, he was startled by how bright the shuttle appeared in what was now a daytime sky. He had watched re-entries before and "I have never seen a light that bright in my life. It was a single contrail but you could see sparkles trailing behind it."

He immediately ran into his house, grabbed a 35-millimeter camera with a 200-millimeter telephoto lens and raced back outside with his 28-year-old son. He began snapping-pictures. Looking through the camera, Dietz saw two large pieces of debris break away.

"I could actually see the change in aspect of it, you could see it tumbling," he said. "The contrail behind it would get wider, narrower, wider, narrower, like that. You can imagine seeing that in real life. I was in shock. I was shaking like a leaf because I knew what was happening."

The shuttle Columbia was breaking up before his eyes in the sky above.

Dallas television stations carried Columbia's re-entry live, providing clear images of the shuttle's destruction. But it took a few moments before the Saturday morning commentators realized what the images meant. Within minutes, what wreckage survived the intense atmospheric heating crashed to Earth along Columbia's flight path, falling vertically after slowing in the denser lower atmosphere. The debris fell like rain.

Investigators later would document debris reports from across central Texas. A motorist outside Lufkin nearly wrecked when a piece of the shuttle hit her windshield. Spooked cattle stampeded. A fisherman saw a fragment hit the water in Toledo Bend Reservoir. Calls reporting smoking debris and loud booms poured in to 911 operators. It didn't take long for authorities to connect the events to Columbia's disappearance.

In Mission Control, televisions were not tuned into outside channels and the flight control team did not yet know exactly what had happened. All the controllers knew was that communications and data had been lost after a series of puzzling sensor failures. There was growing concern, but some still held out hope of re-establishing communications.

That was about to change.

Ed Garske, an off-duty engineer and Mission Evaluation Room manager, was driving a Cub Scout troop from Houston to Corpus Christi for a weekend camp out on the USS *Lexington,* a World War II aircraft carrier now on permanent display as a museum. He had called Don McCormack 45 minutes earlier to make sure Columbia had started its descent. He knew the shuttle would be flying over central Texas right at 8 a.m. local time and thought the kids might have a chance to see the spacecraft's plasma trail low on the northern horizon.

About 15 miles south of Houston, he pulled over and they got out. And sure enough, there was Columbia. It would take Garske a moment to digest what he was seeing.

In the Mission Evaluation Room, McCormack already knew Columbia was in trouble. Like everyone else, he had assumed the loss of hydraulic sensors in the left wing seven minutes earlier was the result of some common component failing and that the flight control team would figure it all out sooner or later. Then Columbia's flight computers sounded a cockpit alarm and reported they had lost pressure data from both of the shuttle's left main landing gear tires. Seconds later, Husband was cut off in mid-transmission.

Nearby, Rodney Rocha was walking between two rooms, nervously checking data from the shuttle's re-entry systems. He knew a few sensors had dropped out here and there, but nothing that meant an impending catastrophe. Suddenly, an engineer looking at data from Columbia's landing gear system got Rocha's attention: "Rodney, what does that mean?"

"On the screen in a little box was 'main landing gear down lock engaged,' " Rocha recalled later. "I said 'I think that means the main gear on the left side is deployed.' "

Even in the thin air 37 miles up, deploying a landing gear at 12,500 mph would be catastrophic.

"It was a sickening feeling," Rocha said. "I ran over to the mechanism side where the landing gear [displays were] and said, 'The display, find the main landing gear door downlock, find it, find it!' "

The engineer called up the display. Rocha looked at it and said, "You tell me what it means. You tell me what that means."

"He said, 'It means the left main landing gear is deployed, Rodney.' " Rocha was stunned. If the landing gear really was down, if this wasn't a sensor malfunction, Columbia's crew was as good as dead.

He looked up and saw Joyce Seriale-Grush. Tears were streaming down her face.

"Joyce, what's the matter?" Rocha asked.

"We've lost communication with the crew," she said.

The two hugged. Rocha ran outside into the hall and called his wife, waking her up. He said, "Pray for us, Robyn. We've lost communications with the crew. I've got to go."

He hung up and ran back inside.

It was clear by now that something had gone terribly wrong. For McCormack, the only thing that could explain the loss of data from unrelated sensors located throughout the left wing was a breach in the wing itself, some sort of opening that had allowed super-heated air to burn its way inside. Like many others by now, he was thinking about the launch-day foam strike.

Suddenly McCormack's phone rang. It was Garske.

"Don, Don, I saw it!" Garske said breathlessly, his voice taut with emotion. "It broke up!"

"Slow down, Mike. What are you telling me?"

"I saw the orbiter. It broke up." Garske and his wife had been at Kennedy when Challenger exploded. Now they looked at each other, tears welling up. "No, not again," he said. "It can't be."

At about that moment in Mission Control, Phil Engelauf's telephone rang. It was Bryan Austin, an off-duty flight controller who had stepped out of his house to look for Columbia's trail low on the northern horizon.

Austin's voice "was pretty shaky and he said [he was] watching the over-flight and [he] saw multiple objects. He had seen that the vehicle had broken up."

By this point, Engelauf said, the flight control team already knew Columbia must be gone. "When that phone call came in, we had already gotten past the point where we should have received radar," he said. "In my mind, that was the absolute, no kidding, black-and-white end. If the radar is looking and there is nothing coming over the horizon, the vehicle is not there. Everything up until then, your heart wants to hold out just that little bit of hope that we are going to get comm back in just a minute and we are going to find out that this was all OK."

Engelauf hung up the phone and turned to his right, relaying Austin's report to astronaut Ellen Ochoa. She gasped and turned away, a look of horror on her face. Engelauf called to LeRoy Cain, who eventually turned around to face him over Engelauf's console. Cain shook his head, digesting the news. Then he straightened up, turned back to his flight control team, and declared a "contingency," the phrase NASA uses to signal a disaster.

"Lock the doors," Cain ordered, taking the first step in a detailed plan designed to impound the data that had been received from Columbia.

"This is Mission Control, Houston," Hartsfield said over NASA Television. "Flight controllers continue to seek tracking or communications with Columbia through Merritt Island tracking station. Last communication with Columbia was at 8 a.m. Central time, approximately above Texas as it approached the Ken-

nedy Space Center for its landing. Flight director LeRoy Cain is now instructing controllers to get out their contingency procedures and begin to follow those."

Nine hundred miles to the east, the milling spectators and family members awaiting Columbia's arrival at Kennedy were beginning to realize something was wrong. At the north end of the runway, Jerry Ross got word Columbia had not been picked up on radar approaching Florida. He immediately knew what that meant. He stepped out of the convoy commander's van and said a brief prayer. Then he and Bob Cabana started calling the astronaut family escorts at the midfield viewing site.

"We think we've lost the vehicle," Ross told them. "We need to get the families rounded up and send them back to crew quarters as soon as you can. Don't say anything."

He then called ahead to the crew quarters staff, saying "We think we've lost the vehicle. Secure the facility. Get security out there, enhance it. Turn off the TVs. Tell everyone what is happening. We're coming back and the families will be there as soon as we can."

Lani McCool was sitting at the top of the family bleachers, straining to hear Hartsfield's commentary over the loudspeakers. Like Jon Clark, she had heard enough to know Mission Control was working on a problem.

"I looked around and to my left, towards the nonfamily side of the bleachers, I saw men in suits, high-level NASA as I was able to recognize a few, all putting cell phones to their ears. Soon everyone was on the phone. Then they started walking towards the parking lot. I ran down the bleachers to ask one of our family escorts what was going on when someone grabbed my elbow and told me we had to go. The whole event was surreal."

Evelyn Husband and her children, Matthew and Laura, were waiting for the sonic booms that would herald Columbia's arrival. The two distinct booms, caused by the shape of the shuttle's wings, usually arrived about two minutes before touchdown.

"We didn't have a clue what was going on," Evelyn later told *The Amarillo Globe-News*. "We were all standing there at the landing site, very light hearted. Matthew was running around playing."

The seconds ticked by.

"I asked what direction the boom would come from and our [astronaut] assistant had the most horrible look on his face," she said. "Laura asked if everything was all right, and Matthew was absolutely silent."

She called her father on her cell phone "and he was crying," she told the Amarillo newspaper. "I asked him if it looked bad, and he said 'yes.' "

Jon Clark was still thinking about bailout options when the family escorts began rounding everyone up.

"It was like the president's been shot," he recalled. "You know, all these guys are pushing you and getting you on the bus or in those cars as fast as you can to get out of there. And nobody's saying anything, just 'we've gotta go, gotta go, gotta go.' "

Ross, Cabana, veteran shuttle pilot Pam Melroy, and other "blue bags"—astronauts in their blue flight suits—left the landing support convoy. Ann Micklos watched them go, knowing by now that Columbia wouldn't be landing at Kennedy. She knew Dave Brown's parents were in Virginia, watching the re-entry on television. She pulled out her cell phone.

"I called his mom, got through right away. She was asking what was happening, and I said there seems to be a problem, I am not really quite sure what is going on. She said they were watching it on TV and they are not being real clear on what is happening. We chatted a little bit more, and I finally got to the point, 'I just wanted to prepare you that we may have lost the vehicle.' I think by that time, that was the general impression that was beginning to go across the TV screen."

The shuttle's landing time came and went, and the big countdown clock began ticking forward from 9:16 a.m. Word began to ripple through NASA, its contractors, and across the nation as cable news programs and the networks interrupted their Saturday morning shows.

NASA administrator Sean O'Keefe, Bill Readdy, agency counsel Paul Pastorek and Kennedy director Roy Bridges quickly left the runway and drove back to the Launch Control Center in Bridges' car. Launch director Mike Leinbach, Mike Wetmore and Wayne Hale headed back in a government van. They all met up in Leinbach's office.

Readdy never went to a launch or landing without a copy of NASA's contingency plan, a detailed set of procedures to guide managers through the process of dealing with a launch or landing mishap. One of the first items on the list was to call the White House and inform the president.

O'Keefe went into a small conference room and called White House Chief of Staff Andrew Card at 9:29 a.m. Card had been following Columbia's re-entry on the NASA television channel and already knew the shuttle was down. Card told O'Keefe that President Bush was at Camp David, jogging, and to call back in 10 minutes or so.

"So it was, like, 9:45 or so when I talked to him," O'Keefe recalled, "and I said, 'Mr. President, we had an accident today. It's a disaster.' And his first response was 'Where are the families? Are you taking care of them? Make sure they are being kept away from the press folks and everything else and that their privacy is being respected.'

"His next observation was 'Keep me posted, tell me what it is as you know it, let me know what's going on. I would like to make a statement later today. Let's think about how you want to do that because I think you ought to make yours first and then we will do something a little later on. But we have to tell everybody what it is we know as soon as we know it.' "

O'Keefe then called Homeland Security Secretary Tom Ridge and Stephen Hadley, the White House deputy national security advisor. They discussed the possibility of terrorism briefly, but it soon became obvious the shuttle's extreme altitude and velocity made that virtually impossible.

Leinbach and several others were thinking about the launch-day foam strike. He ordered copies of a photo showing the spray of white debris emerging from under Columbia's left wing, and Readdy told him to pass them around.

"So I gave a copy of the picture to everyone in the room, including Mr. O'Keefe, not because I was sure that was what it was, but at that time it's the only thing I could think of that could have been [the cause]," Leinbach recalled.

In Houston, Linda Ham didn't dwell on what might have caused the disaster. She, too, thought about the foam strike, but she had more immediate work to do. She called Dave Whittle, predesignated chairman of NASA's standing Mishap Investigation Team, a group of engineers and managers on call during every launch and landing, ready to swing into action in case of an emergency. Whittle had written many of the procedures in the contingency plan Readdy and the others were implementing.

Ham made the call from the manager's room overlooking Mission Control.

"It was pretty quiet in there," she said. "There was no radar. It was after landing time, and it was so quiet up there. I said to Ron, 'I need to call Dave Whittle,' because I'm thinking we just can't stand here and not do anything."

The contingency plan called for a Mishap Response Team meeting one hour after the plan was activated. O'Keefe had called the White House around 9:30 a.m. So Ham scheduled the meeting for 10:30 a.m. She would chair the discussion in Houston, and the teams at Kennedy and Marshall would participate via teleconference.

Ham left Mission Control and walked across the Johnson campus to her office in Building 1 to pick up her own copy of the contingency plan. For some reason, she had not brought it with her that morning. Before heading back, she took a minute to phone home.

"My son, Ryan, answered and said that Dad was back and in the barn with the 9-year-olds, coaching baseball practice," Ham said, her eyes filling with tears as she recalled the moment. "So I said to Ryan, 'The shuttle isn't going to land because it came apart over Texas.' He asked me what happened, and I said, 'We really don't know what happened. We're going to be having meetings all day and night, and I don't know when I'll be home today.' And he says, 'That's OK, Mom. You can tell us about it when you get home.' "

Back at Kennedy, someone had turned on a television and O'Keefe and the others watched in silence as Columbia broke apart again and again in endless replays. It was the first time they had actually seen the orbiter's destruction.

"We were just stunned, absolutely stunned," Leinbach said. "I remember it like it was yesterday. Mr. O'Keefe put his hands on the table, like this"—he stood, putting his hands on a table, head and shoulders sagging—"and said, 'I wonder how many people on the ground were just hurt?' I had no idea where it was, other than east Texas. So we started talking about the contingency plan and getting the telephone calls made."

O'Keefe remembered, "[I] was convinced that at some point during the course of the day, any minute now, we were going to start seeing reports of big pieces of the orbiter blasting into buildings and homes and killing somebody."

Micklos and the other engineers and technicians in the landing convoy had made their way back to their original staging area by the shuttle runway to await instructions. No one knew what to do. Someone led a prayer for the crew.

Micklos raised her hand and asked if she could speak.

"I identified myself, said that I had dated one of the astronauts, Dave Brown, and I said we had a conversation, if something should happen to the orbiter, what would you like me to convey to everyone?" she told the group. "With that, I spoke his words, which were that he wanted me to make sure that I could find the people or person that may have been the cause of the accident."

"I want you to go up there and tell them that I hold no animosity for what happened," Brown had told her. "I died doing what I loved and have no regrets."

The group listened in silence, many overcome with emotion. Even grizzled shuttle veterans choked back tears.

The Mishap Response Team meeting began at 10:30 a.m. and lasted less than two hours. It was obvious that a large team would need to reach the crash site, wherever that was, as soon as possible to identify debris and potentially dangerous components. Explosives were used throughout the spacecraft, along with extremely toxic rocket fuel and other materials. If any of that made it to the ground intact, whoever stumbled across it would be in danger.

The top priority was recovering remains of the crew. Jim Wetherbee, commander of the Endeavour mission the previous November, and other members of the astronaut office at Johnson set up that phase of the operation. Out of respect for Husband's crew and their families, the search would be carried out in strict secrecy.

The question was where to begin. Whittle settled on Barksdale Air Force Base in Shreveport, Louisiana, as his initial center of operations. The shuttle broke apart over eastern Texas, but there were no communities along its flight path with airport runways long enough to support the big transport planes that would be required. Kennedy's Rapid Response Team, chaired by Leinbach, planned to depart that evening aboard a transport plane that had been flown in from Charleston, South Carolina.

When the Mishap Response Team meeting broke up, Leinbach's rapid response crew started making calls to get luggage delivered, to gather needed equipment, and to make arrangements for housing and cars in Shreveport.

The astronaut families had arrived at crew quarters around 9:30 a.m. Ross and Cabana had gotten there first and had already seen video of Columbia disintegrating above Texas. Ross was waiting at the elevator, and he led Evelyn Husband and the others into a conference room to await word on what had happened.

"I think when I escorted her from the elevator to the conference room, she knew," Ross said of Husband. "She was trying to console her kids, but she was going to hold on to the last possible minute as far as what she really admitted to."

Ross walked to Cabana's nearby office and told him the families had arrived.

The mood among the families was "horrific," Clark said. A sense of foreboding had settled in. Making eye contact with members of the crew quarters support staff, Clark saw his worst fears confirmed. The children did not immediately understand: "They were just like, 'What's going on?' You know, 'Where's mommy and daddy?' And then Bob comes in and he talks to us."

The ex-Marine fighter pilot and shuttle commander told the families Mission Control had not picked up any signals from radio beacons that would have activated if the astronauts had managed to bail out. He told them that at Columbia's altitude and velocity, it was not likely anyone could have survived.

"I thought he was pretty brave, because he obviously didn't have a lot of insights," Clark said. "He said, 'We lost contact with the vehicle.

It was at an altitude and an airspeed that was probably not survivable. We don't have confirmation of that, but I don't want to give you a false sense that they're OK.' And then everybody just breaks out in tears and crying, and the kids, some of them were screaming. It was horrible. And then everybody just starts hugging and crying. And the crew quarters staff comes in, the family escorts are there, everybody's just crying."

Cabana remembered Mike Anderson's kids "turning to their mom and just exploding in tears and hugging her, one on each side."

And that set Iain off as well.

"They started screaming, they were about his age, and then Iain started to scream. But then he got over it," Clark said, "and he was crying, and then somebody had the foresight to take all the young kids down the hall to another little lounge there. Barbara Morgan was there. She's been through this [as a back-up astronaut in the Challenger disaster]. I remember talking to her about it right afterwards. I said, 'This must be horrible for you.' And she was so calm, she knew what to do with the kids. She got them in there, and I mean 30 minutes later, they're just playing, some are watching TV or playing video games or board games.

"And Iain's in there, and I'm like, you know, a wreck. Iain said something to me. He said, 'You're bumming me out because you're so sad.' And I said, 'Well, I'm sorry. You're going to have to help me through this.' "

Ross stayed with the families a few minutes and then escorted Rona Ramon and her children to Ilan's bedroom so she could call home to family in Israel.

"Then I went back down, started running the rest of the operation," Ross said. "I had to find a way to get them home."

The families had intended to spend the night in the Cocoa Beach area while four of the astronauts underwent detailed physical exams and post-landing medical tests as part of the mission's research. They were not prepared to fly back to Houston. Room keys were collected, and support teams visited each family's motel room, packing their bags for them, paying the bills, and taking the luggage to the nearby Cape Canaveral Air Force Station runway. Others called local police departments around Johnson, asking for officers to stand by outside each family's home. Ross asked Kennedy officials to provide an executive jet to help ferry the families home. It would join two jets from Johnson that were already on standby.

O'Keefe left the Launch Control Center around 11:20 a.m. and headed for crew quarters. He spent a quarter of an hour consoling family members before calling the White House to arrange the presidential conference call. Bush offered the families his sympathy, saying the nation appreciated their sacrifice. He said the astronauts had dedicated their lives to a noble objective and that a grateful nation would never forget.

"I was numb," said Lani McCool. "I know he said something about prayers and their hearts being with us."

Food was brought in. Close friends and extended family members, who had been taken to a nearby auditorium, were ushered in for brief visits. Less than three hours had passed since Columbia's destruction.

At NASA headquarters in Washington, Deputy Administrator Fred Gregory, another former shuttle commander and Bryan O'Connor's predecessor as chief of shuttle safety, was working through his part of the contingency plan. The document included procedures for getting a disaster investigation board up and running immediately.

The contingency plan specified a half-dozen board members, but it did not name a chairman. Former Navy Secretary O'Keefe thought of two possibilities: Joe Lopez, a retired admiral who had served on a number of military accident boards, and Hal Gehman, another retired admiral who had led one of three probes into the October 2000 terrorist bombing of the USS Cole, a guided missile cruiser docked in Yemen. O'Keefe asked Gregory to call both men. Lopez had other commitments, "but Joe was extremely positive about [Gehman]," O'Keefe recalled later.

Gehman had no connection with the space program, did not know Gregory, and barely remembered O'Keefe from a chance encounter years before. Gehman had spent that morning in a lawyer's office in Williamsburg, Virginia, reviewing his father's financial affairs. His brother arrived late, around 10 a.m., and told the group the shuttle had crashed. They went on with the family business.

As he was driving back to Norfolk, Gehman's cell phone suddenly beeped. He had been out of range and discovered he had several messages. One of them was from Gregory.

"It was absolutely out of the blue," Gehman said. "I was rolling down the highway with my parents in the car. I called Mr. Gregory, and he said that Sean O'Keefe was just closing up business in Florida, was about to get on the airplane and fly back to D.C., but he wanted to ask me if I would serve as chairman of the predesignated accident investigation board."

Gehman told Gregory he had no experience with shuttle operations and was not a pilot. Gregory replied, "Nope, we are not looking for an aviator. We are looking for an investigator, and what it says here is that you did the investigation into the Cole, which was highly regarded, and that is what we are looking for."

Reassured the White House had approved the assignment, Gehman agreed, dropped his parents off, and drove home to tell his wife. "She said, 'What do you know about space shuttles?' And I said, 'I don't know anything about space shuttles. But maybe that's the right attitude.' "

O'Keefe, meanwhile, had another quick chat with the president to bring him up to date. It was decided the administrator would hold a news conference at 1 p.m., NASA's first public statement since the disaster. The president would fly back to the White House and address the nation at 2 p.m.

Ross and his team got the families ready for the trip back to Houston, driving them in a convoy to the Cape Canaveral Air Force Station runway where three jets were waiting.

"After I put the guys on the airplanes, as I was driving back across the causeway, I had to pull over to the side for a while," Ross said. "I just lost it. I was able to hold it together pretty well until then because I flat had too much to worry about. So I spent a couple of minutes there thinking about my friends, saying some prayers, and then got back on with it."

O'Keefe headed for the Kennedy press site just across the street from the Launch Control Center, where NASA public affairs was struggling to cope with jammed phone lines and endless questions from reporters. He called the Launch Control Center for an update and then joined Readdy in the press site auditorium to hold NASA's first news conference.

"This is indeed a tragic day for the NASA family, for the families of the astronauts who flew on STS-107, and likewise is tragic for the nation," O'Keefe began. He briefly mentioned the implementation of the contingency action plan, his conversations with the president and the secretary of Homeland Security, and his visit with the family members.

"The president has called and spoken to the family members to express our deepest national regrets," O'Keefe said. "We have assured them that we will begin the process immediately to recover their loved ones and understand the cause of this tragedy."

He said NASA was forming an independent accident investigation board and urged the public not to handle any suspected shuttle debris.

"The loss of this valued crew is something we will never be able to get over," he concluded. "We have assured the families that we will do everything, everything we can possibly do to guarantee that we work our way through this horrific tragedy. We ask the members of the media to honor that too. Please respect their privacy and please understand the tragedy that they are going through at this time."

As O'Keefe was talking to reporters, four astronauts were departing Houston on a three-hour drive to Lufkin, Texas, where initial debris reports were beginning to flood in. At that point, Wetherbee said, the best estimate NASA had was that wreckage might have fallen along a 50-mile-wide swath stretching 250 miles across eastern Texas.

Wetherbee and the other three astronauts had no idea what to expect. They took spare batteries, maps, flashlights, cell phones, and Global Positioning System receivers to help pinpoint the locations of anything they found.

"We knew we were going to have tremendous difficulties," Wetherbee said. "We knew that we had a responsibility to the immediate family members . . . to recover the remains with dignity, honor, and reverence and to do so as quickly as we could. Everybody was offering to help. The difficulty is trying to figure out what to do when you have no rules and you are trying to go find seven of your friends and you have to start an organization like that."

Wetherbee reached Lufkin late in the day, after stopping briefly to inspect reported shuttle debris, and met with Marshall deputy director Dave King. O'Keefe had dispatched King to Texas to coordinate initial search efforts. They moved NASA's command post from the local FBI office to the Lufkin Civic Center.

Within a few hours of Columbia's destruction, Bush had declared east Texas a federal disaster area, allowing the Federal Emergency Management Agency and the Environmental Protection Agency to send teams to the debris field to coordinate recovery operations. FBI Evidence Recovery Teams flew in who were veterans of the Oklahoma City bombing and the World Trade Center reconstruction, along with agents from Dallas and elsewhere. Hundreds of state and local police officers and emergency personnel began fanning out across the debris zone.

By Saturday afternoon, the National Guard had been called up to help protect areas where debris had been found. The area's woods and pine forests were filled with law enforcement officers and local residents working side by side to recover any pieces that had fallen from the sky. Remarkably, there were no reports of serious injury. Several people had close calls. A dentist reported a foot-long piece of debris smashed through the roof of his office. An airport worker said a spherical tank crashed onto a runway. In Nacogdoches, a piece of Columbia hit between two tanks of explosive natural gas positioned just a few feet apart. But no one on the ground had been hurt.

As initial recovery efforts got underway, President Bush addressed the nation from the White House.

"My fellow Americans, this day has brought terrible news and great sadness to our country," the president said. "At 9:00 a.m. this morning, Mission Control in Houston lost contact with our Space Shuttle Columbia. A short time later, debris was seen falling from the skies above Texas. The Columbia is lost; there are no survivors."

He mentioned each astronaut by name and praised them for assuming "great risk in the service of all humanity."

"In an age when space flight has come to seem almost routine, it is easy to overlook the dangers of travel by rocket, and the difficulties of navigating the fierce outer atmosphere of the Earth. These astronauts knew the dangers, and they faced them willingly, knowing they had a high and noble purpose in life. Because of their courage and daring and idealism, we will miss them all the more."

Once again, he offered comfort to the families, already on their way home to Houston, saying, "Those you loved will always have the respect and gratitude of this country."

"The cause in which they died will continue," Bush added. "Mankind is led into the darkness beyond our world by the inspiration of discovery and the longing to understand. Our journey into space will go on. . . . The same Creator who names the stars also knows the names of the seven souls we mourn today. The crew of the shuttle Columbia did not return safely to Earth; yet we can pray that all are safely home.

"May God bless the grieving families, and may God continue to bless America."

At this point, NASA had not discussed what might have gone wrong. The Mission Control discussions about left wing sensor failures were not carried on the NASA television channel. All that was known from an official standpoint was that communications had been lost and Columbia had crashed.

Word was nonetheless beginning to trickle out. Within the first hour after the accident, at least two reporters heard about the hydraulic system sensor failures and reported that whatever had happened, it had started in the left wing. Other journalists found out as well, but in the absence of anything official, the public waited for word from NASA. It came at 3 p.m.

Ron Dittemore and Milt Heflin, chief of the flight director's office at Johnson, held a news conference to explain what little was known up to that point.

"I'm sure you understand how difficult a time this is for us right now," Dittemore began. "We're devastated, because of the events that unfolded this morning. There's a certain amount of shock in our system because we have suffered the loss of seven family members, and we're learning to deal with that. As difficult as this is for us to do, we wanted to meet with you and be as fair and open with you, given the facts as we understand them today."

He warned the public to be careful handling debris and said, "At this stage, I haven't received any real information on debris or status of crew remains."

The emotional trauma of the day was apparent in both men's faces.

Heflin looked particularly hard hit. He was a veteran of the Challenger disaster and, before that, the launchpad fire that killed three Apollo astronauts in 1967.

"Sometimes it's a shame that it takes things like this for this country to pull together and care," he said. "And it shouldn't. Man, we're good. This country is great. It shouldn't take these kinds of things to cause a coming together."

Heflin then walked through the final bits of data from Columbia: how the hydraulic sensors had stopped working, how temperatures began rising in the left main landing gear wheel well, how flight controllers lost data from temperature sensors on the skin of the shuttle. He told reporters about the loss of tire pressure data that triggered a cockpit alarm. He said Columbia was 207,135 feet up and traveling at 18.3 times the speed of sound when communications were lost.

Dittemore was asked about the launch-day foam impact. In the hours immediately after the mishap, word about the foam strike spread through the ranks of reporters. Those who had heard about it earlier dug out their notes and began reporting it for the first time.

But Dittemore downplayed the strike, saying, "We spent a goodly amount of time reviewing that film and then analyzing what that potential impact of debris on the wing might do. Through analysis and through our ability to call back on our experience with tile, it was judged that the event did not represent a safety concern."

He agreed there might be a connection between the foam hit and whatever caused the sensor failures during Columbia's re-entry because both involved the left wing.

"But we have to caution you and ourselves that we can't rush to judgment on it because there are a lot of things in this business that look like the smoking gun but turn out not even to be close," he said.

The questions continued. One reporter asked if NASA ever considered ordering a spacewalk to look for signs of damage. Dittemore pointed out Columbia had not been equipped with a robot arm and, therefore, the crew had had no way to carry out a remote inspection. Without an arm, there is was "no capability to go over the side of the vehicle and go underneath the vehicle and look for an area of distress," he said.

And it wouldn't have done any good anyway, Dittemore continued, because the astronauts had no tools or other materials on board to repair tile damage.

"We don't believe, at this point, that the impact of that [foam] debris on the tile was the cause of our problem," he said. "We convinced ourselves, as we analyzed it 10 days ago, that it was not going to represent a safety issue. Now, we had the events of this morning. We're going to go back and see if there's a connection.

"Is that the smoking gun? It is not. We don't know enough about it. A lot more analysis and evidence needs to come to the table. So it's not fair to represent the tile damage as the source. It's just something we need to go look at."

Another reporter asked why NASA never asked the Defense Department for close-up photos of the wing, either using ground cameras or spy satellites, to take a look at the impact site. Dittemore said NASA's experience with such photos showed they were of marginal use.

"Combine that—our feeling that we didn't believe the pictures would be very useful to us—with the fact that there was not much, there was zero that we could do about it [to make repairs] and in this case, we elected not even to take the pictures," he said. "We believed that our technical analysis was sufficient. We couldn't do anything about it anyway."

That central belief—"we couldn't do anything about it anyway"—was part of the same rationale used by Ham only 10 days earlier to shut down the requests for photos. That belief later would be sharply criticized by astronauts and outside investigators who would come to believe it might have been possible to save Columbia's crew if the management team had recognized the potential severity of the foam strike sooner.

But at the time, Dittemore's statement reflected the views of many, both in and out of NASA, and reporters generally gave him the benefit of the doubt.

Gehman spent the afternoon at his home, reading contingency plan documents being faxed to him by Gregory's office. The first thing he decided to do was change the name of the accident board. The NASA title was International Space Station and Space Shuttle Mishap Interagency Investigations Board. It soon became simply the Columbia Accident Investigation Board, or CAIB.

The board had six predesignated non-NASA members, all of them experts in aerospace safety and aviation crash investigations. Rear Adm. Stephen Turcotte was commander of the U.S. Naval Safety Center in Norfolk. Maj. Gen. John Barry, director of plans and programs at Materiel Command, was based at Wright-Patterson Air Force Base, Ohio. Maj. Gen. Kenneth Hess was chief of safety at Kirtland Air Force Base.

Providing an academic's perspective was Dr. James N. Hallock, chief of the Aviation Safety Division at the U.S. Department of Transportation in Cambridge, Massachusetts. Also on board were Steven Wallace, director of accident investigations for the Federal Aviation Administration in Washington, and Brig. Gen. Duane Deal, commander of the 21st Space Wing at Peterson Air Force Base, Colorado.

Deal and Barry replaced two other Air force generals who were unable to serve.

Earlier that morning, a seventh member was added to the board: NASA's own Scott Hubbard, director of the agency's Ames Research Center. Hubbard led a major effort at NASA to redesign its Mars exploration program following back-to-back failures in 1999.

Hubbard realized the assignment would raise questions about the accident board's independence from NASA.

"This was a national tragedy, and I wanted to do the best I could to serve the nation," he said. "But I also immediately thought about being the only NASA board member and said to myself, well if I'm going to serve in this role, I'm just going to have to call them like I see them."

O'Connor was named to the board as a non-voting ex-officio member, and Theron Bradley, NASA's chief engineer in Washington, would serve as secretary.

Back at Johnson, flight controller Jeff Kling finally left Mission Control. Sitting in his pickup truck, he checked his voice mail and sobbed as he listened to messages from his children saying "we are behind you" and "we are sorry for you, Dad."

"I sat out in my truck for about 10 minutes to recompose myself and make sure that I could go home," he recalled. He lived in a nearby subdivision called Polly Ranch that featured an airstrip. Many of the homes had hangars and several astronauts lived there, including Columbia's Dave Brown.

As Kling drove into the neighborhood, he was overwhelmed again. "My wife had ordered some flags from the Boy Scouts, had them put out at the entrance to the neighborhood," he said. "They already had a makeshift memorial out there and they had six American flags and one Israeli flag. I only had a block to drive and I didn't know if I could do it. So I got home and we ended up opening our house and having kind of a 'happy Dave time,' sort of a memorial sort of thing. But it was mostly so everybody could kind of grieve together."

He paused for a moment, recalling the day Brown closed on the purchase of his house. The astronaut drove his Honda Civic into his hangar "and he was driving around in circles inside, honking the horn in this big empty hangar, just like a kid. He said, 'I always wanted to live on an air strip and this is just so cool!' "

O'Keefe, Readdy and other members of the headquarters contingent flew back to Washington that afternoon. As associate administrator for space flight, Readdy was the man with overall responsibility for the shuttle program. He offered his resignation to O'Keefe.

"He said, 'This happened on my watch, my responsibility, I am prepared to resign right now,' " O'Keefe recalled. "I said, 'Not accepted. We are going to have to [work] our way through this together, buddy.' "

Back at headquarters, O'Keefe chaired a 5 p.m. teleconference with Gehman and a few of the other board members, telling them to head for Barksdale Air Force Base where the Mishap Investigation Team was setting up shop. A government plane would pick them up on Sunday.

At Kennedy, the Rapid Response Team was finally ready to roll. About 40 engineers and technicians, both NASA civil servants and contractors, loaded their luggage, kissed their spouses good-bye and flew off into the night. The mood in the aircraft was quiet. Many had been up since dawn or before. They were heading west toward the shattered remains of a spacecraft many of them had spent their careers preparing for flight.

They reached Barksdale in the dead of night, walking off the plane into an darkened hangar. They didn't have rooms or rental cars. They didn't know where the debris was or what they were supposed to do next. A sense of melancholy settled in. But Whittle was already there, and soon the group got organized.

"Mike, the first thing I need from you all tomorrow morning is a list of all hazards on board the orbiter so we can start distributing that to the folks in the field, so they know what to expect, what to touch, what not to touch. That kind of stuff," Whittle told Leinbach.

Hours later, Leinbach and the others checked into hastily-booked motel rooms and tried to put the day behind them.

Leinbach turned on the television.

"I was still wound up when I got back to my room," he said. "It was like, 2 or 3 in the morning. I turned on the TV just to try to wind down and there it was again. I just lost it."

EIGHT

AFTERMATH

It's difficult for us to believe, as engineers, as management and as a team, that this particular piece of foam debris shedding from the tank represented a safety of flight issue.

—*Shuttle program manager Ron Dittemore*

It was an unearthly sound, a crackling, ground-shaking roar that rattled windows and shook pictures off the wall.

Roger Coday, a 59-year-old chemical engineer and Vietnam veteran, instinctively dropped to the floor of his mobile home near Hemphill, Texas, as his wife, Jeannie, screamed from a bedroom, "What was that?" The rumbling continued and Coday ran outside. A neighbor had problems with a propane tank the day before, and Coday thought it might have blown up. But when he looked toward his neighbor's house a hundred yards away, he saw the tank was intact. The roaring sound lasted more than a minute. He recalled, "[It was] just a crescendo, just a continuous sonic boom. . . . I ran on around the corner of the house and was facing northwest over an open pasture that's across the road in front of our place. We're in a rural area out in the country. And then I saw the vapor trails, the contrails, white vapor trails coming across the sky."

Coday's brother, Frank, had called from his house four miles down the road. "What was that?" he asked. The two men debated what might have caused the roar and speculated that perhaps a pipeline two miles away to the northwest had exploded. Coday's wife, meanwhile, had turned on a television. She immediately heard NASA had lost contact with the shuttle Columbia.

"It was then that we began to kind of relate the noise to something happening to Columbia," Coday said.

But he didn't immediately make the connection between NASA's loss of contact with Columbia and the smoke trails arcing across the sky. The excitement died down and 20 minutes later, Frank drove over to help with some work around Coday's house, a weekend retreat the couple visited as often as possible. He told his brother, "There's something right up there on the road at the corner to your property." Coday told Jeannie they would be back in a moment, jumped in Frank's pickup truck and took off to give whatever it was a closer look.

"When we pulled up and I got out of the truck and walked down the road, I knew that it was human remains," Coday said.

Another neighbor and his wife pulled up, and then another. The five were standing there, discussing the grisly find, when Hemphill Police Chief Roger McBride, alerted by a 911 call, arrived 10 minutes later.

"He walked up and looked, and he said, 'Yes, I believe that's human remains.' Roger ran back to his car and got some yellow crime scene tape, and we helped him barricade the area off," Coday said.

By now, Texas Department of Public Safety troopers had arrived. McBride knew NASA was dispatching a helicopter to the scene, and

Frank took the truck back to his brother's home to pick up a portable Global Positioning System satellite receiver to pinpoint the exact location of the discovery.

Coday looked around again, spotting broken branches in the dense canopy above: "It began to sink in that this may be the remains of one of the astronauts."

Soon, the NASA helicopter swooped overhead. Astronaut Mark Kelly had arrived from Johnson to oversee the recovery effort. The road was blocked off, and Coday and the neighbors were asked to leave. They went back to the mobile home. Shortly after noon, John Starr Jr., the local funeral director, stopped by and asked for directions to the scene. Later that afternoon, the Codays saw Starr and Kelly drive by in a hearse, heading back into town.

"The astronauts have always been my heroes," Coday said. "As the astronauts came and went, the Mercury program, the Apollo program, I always felt an attachment to it. Then later, when we had the first disaster, all those things just really hit home. But this one, this one was so close . . ."

Sitting on his porch late that afternoon, Coday decided to build a simple memorial. He fashioned a cross out of two cedar logs he had cut earlier and erected it at the site where the astronaut had fallen to Earth.

"It's there and we still maintain it," he said eight months after the disaster, still wondering which astronaut it was. "I am a very devout Christian, and I prayed for that person's soul."

Over the next five days, search teams located remains of all seven shuttle fliers within a few miles of Coday's home, the same area where Columbia's reinforced carbon-carbon nose cap would be found crushed and partially buried in the Texas soil.

While a few crew cabin items were filmed by news photographers in the immediate aftermath of the disaster—a battered pressure suit helmet, a cloth mission logo patch and a handful of others—news accounts and photographs quickly dried up as search teams got better organized and NASA clamped down on the release of details regarding the crew's fate.

Jim Wetherbee would not discuss crew recovery operations or confirm when or where the astronauts were found. But he said the swift recovery of Columbia's crew was the result of an exhaustive around-the-clock effort by hundreds of federal, state, and local workers.

"The Federal Emergency Management Agency came in the next day, and we had the Environmental Protection Agency helping, the Texas Department of Public Safety, Texas police officers, fire departments, local and state, Texas Forestry Service, Texas Army National Guard, various medical groups, recovery forces, rescue forces, contractor companies, Salvation Army, local businesses and the citizens of the whole country and in particular, east Texas," Wetherbee said. "We formed the search teams relatively quickly. We did find the remains of the seven crew members, ceremonial last rights were administered with reverends in the field, and we escorted the remains to their resting place."

On Feb. 5, remains of all seven astronauts were flown to Dover Air Force Base in Delaware for DNA tests and positive identification at the Charles C. Carson Center for Mortuary Affairs, the same facility that prepared Challenger's crew for burial. Deputy Administrator Fred Gregory stood by on the tarmac near an honor guard as the flag-draped coffins were carried off an Air Force transport jet one by one.

It was a painful moment for NASA. But everyone involved felt a sense of relief that the recovery operation had proceeded so swiftly. When Challenger went down off the coast of Florida 17 years earlier, it took months to recover identifiable remains of all seven crew members. By Feb. 13, Columbia's astronauts had been positively identified, and the remains were released to the families shortly thereafter.

"We are comforted by the knowledge we have brought our seven friends home," Bob Cabana said in a statement. "We are deeply indebted to the communities and volunteers who made this homecoming possible and brought peace of mind to the crew's families, and the entire NASA family."

Early Sunday morning, 24 hours after Columbia fell to Earth, thousands of mourners in the communities surrounding the Johnson Space Center filed into area churches to remember the astronauts, to honor their life work and to pray for their families. The interdenominational

Grace Community Church in Clear Lake, where Husband and Anderson were active members, was packed with friends and supporters, along with dozens of reporters, photographers and television camera crews, who crowded inside for hymns, prayers and remembrance.

Pastor Steve Riggle had heard about the disaster the day before in Guatemala, where he had been visiting a church sponsored by Grace. He flew home that night to face a devastated congregation.

"It's incredible that one moment changes everything," he said from the pulpit. "It's the difference between tragedy and triumph. A mere 16 minutes from what would have been considered great victory. But instead we're gathered today around tragedy. Every one of us in this room [is] affected by it in some way. That's why we're here."

At nearby Johnson, flags were flying at half-staff. At the main entrance to the space center, surrounding a large sign facing NASA Road 1, flowers, banners and photographs of the astronauts materialized overnight as grieving citizens and area space workers created an impromptu memorial to Columbia and its crew. In the days ahead, the colorful memorial, with thousands of flowers and handmade signs and banners, would grow to the point that police would be forced to close one lane of traffic on the busy highway to provide parking for the steady flow of visitors.

Just inside the gates to the space center, the grounds around Johnson's public affairs office and central auditorium were crowded with satellite television trucks and news crews that had raced in the day before or driven through the night to reach Houston from towns and cities across the nation's heartland. Television correspondents were scattered about, participating in Sunday morning talk shows, while print and radio reporters mobbed Johnson's public affairs office seeking interviews, background material and answers to technical questions. For real answers, the media would have to wait until a news conference that afternoon.

While the families and the public mourned, Harold Gehman's Columbia Accident Investigation Board and a huge team of NASA and contractor engineers kicked off an around-the-clock effort to find out what had gone wrong.

It was a daunting task. The spacecraft broke apart 37 miles up, traveling 18 times the speed of sound, and its shredded wreckage was spread across a vast area. There were no crash-resistant "black boxes" like those aboard commercial airliners. The voice and data recorders aboard Columbia were not designed to withstand a high-altitude, high-speed breakup, and no one knew if any had survived the plunge to Earth. As far as the investigators knew, they would have to piece together the cause of the mishap based primarily on the condition of recovered wreckage and data radioed to Earth before the shuttle disintegrated.

It would not be easy, and it would not be quick. But there was a sense of urgency in the air. The space station's three-man crew was nearing the end of its normal tour of duty. NASA had planned to launch a replacement crew aboard the shuttle Atlantis a month after Columbia's landing to bring Ken Bowersox, Don Pettit, and Nikolai Budarin back to Earth after three and a half months in space. With the shuttle fleet grounded, it was clear to everyone Bowersox and company would have to remain in orbit longer than planned.

In the meantime, assembly of the multi-billion-dollar outpost was on hold, and without periodic shuttle resupply missions, the long-term fate of the lab was uncertain. By coincidence, the Russians launched an unmanned Progress supply craft from Kazakhstan on the day after the disaster. It was carrying equipment, fresh food, and supplies for the current fliers and their replacements. With the shuttle out of action, it would be up to the Russians to keep the station in operation. But Progress vehicles held only a portion of the supplies carried by a shuttle and not nearly as much fresh water as was generated in orbit by the shuttle's fuel cells. Without shuttle visits, it was not clear the station could continue to support a three-person crew.

Gehman wasn't one to waste time. As churches were letting out in Clear Lake, board members were making their way to Barksdale Air Force Base.

Aboard one government jet, Gehman and Steve Wallace began to get acquainted. The FAA official was impressed.

"Gehman came back and we sat around the table and that was my first real introduction to him," Wallace recalled. "He asked several of us our thoughts about the investigation, a little bit about ourselves, and about what we thought about this being a lot of work. . . . My first impression was that he was actively listening to each one of us, and it was confirmed the next day when he was sort of repeating back what various others had said to him in a larger group. It was a very good start to the whole thing. It never faded."

That afternoon, the board members and staff who got to Barksdale early held an informal discussion prior to the first official meeting of the panel later that evening. Most of that initial discussion centered on administrative issues, but hints of the board's general attitude can be gleaned from the official minutes of the meeting.

Gregory told the board White House interest in the investigation was high and that the administration intended to closely follow the progress of the probe. He also mentioned NASA was actively resisting "the push for a presidentially appointed or congressionally appointed commission" like the one that investigated the Challenger disaster and that Congress was generally "supportive of the Gehman Board, at least until it gets screwed up." The minutes didn't reflect the widely held perception the Bush administration did not consider the space program a top priority.

Board member John Barry discussed the need for openness while still retaining an ability to obtain "candid and sensitive input from NASA and contractor employees; this is not a public witch hunt." Bryan O'Connor responded that NASA did not typically use sworn statements in accident investigations because they can "imply legal culpability." Instead, he said, the agency preferred the use of "witness statements" that make it clear "we are solely interested in fact finding." Wallace said FAA witness statements seemed similar to NASA's.

Other board members arrived at Barksdale later that afternoon, and the panel held its first official meeting that evening. Board members understood from the outset that they faced contradictory pressures. According to the minutes of the meeting, it was decided "the overall schedule is driven by the need to support ongoing manning of the International Space Station (go fast) and ultimately by the need to make the shuttle safer (go slow)." The board's initial charter called for a final report within 60 days of the accident; "but it was recognized that the Board would be done when it got done," the minutes noted.

Back at Johnson, reporters had received word a large piece of Columbia's forward fuselage had been located. Remembering the way

Challenger broke apart in 1986, veteran reporters speculated Columbia's reinforced crew module might have made it to the ground roughly intact, or at least with its wreckage concentrated in a relatively small area. As it turned out, the latter scenario was essentially correct. Columbia's nose cap was quickly found, along with other debris from the forward fuselage.

But the big question remained: What brought the shuttle down? At an afternoon news conference, Ron Dittemore painted a tantalizing picture of a doomed spacecraft struggling to stay on course as an unknown problem of some sort steadily worsened, pulling the shuttle's nose to one side like a car with misaligned wheels.

It was a remarkably detailed presentation, one that stood in stark contrast to a strict "no comment" policy implemented in the wake of the Challenger disaster. In that case, senior astronauts and managers in charge of the shuttle program decided not to say anything about what might have caused the mishap until the presidential commission investigating the accident issued its final report. The policy was carried to such extremes that public affairs representatives were not allowed to use the words "fire" or "flame" when describing film showing a jet of burning propellant shooting out of Challenger's right-side booster. They were forced to call the jet an "anomalous plume." Relations with the press soured.

The resulting antagonism had been widely viewed as another sort of disaster for NASA, one that cost the agency much goodwill. Dittemore said the decision this time around to hold daily news briefings and to pass along technical information wasn't so much a conscious policy change as a desire on his part to keep the space community informed about the progress of the initial investigation.

"My primary motivation was to get out in front and talk so the people would understand what was going on day to day," he said later. "It was not a hard thing for me to do. In fact, I felt like I was doing good for the people. I felt like they knew what was going on, and I was anxious to continue doing that during the first week of grief."

At the news conference, he explained, in answer to a question, "I'm going to be honest and open with you and tell you exactly what we know and hope you understand that from day to day, it will change."

He retold the events of Saturday in Mission Control, from the initial dropout of temperature data from the left wing. He added that at about the same time, temperature sensors on a left main landing gear brake fluid line and the landing gear strut had begun showing an increase.

"This is significant in that these measurements were located in the left wheel well," Dittemore told reporters. "This was the first occurrence of a significant thermal event."

Somehow, heat was getting into or very near the wheel well. The wiring to the hydraulic sensors at the back of the wing was routed along the outboard and forward sides of the wheel well box. A major problem there could have affected that wiring and interrupted the flow of data from the other sensors.

The next hint of unusual heating had come from the shuttle's left fuselage, well above the wing. Four minutes later, at 8:58 a.m., Columbia was over New Mexico with its nose pointed sharply to the left as planned—but data showed the shuttle's flight computers were now repositioning the elevons at the back of each wing to battle an unexpected aerodynamic drag on the left side of the spacecraft. That drag, which probably was caused by something disrupting the smooth flow of air over the left wing, was trying to pull Columbia's nose farther to the left. But it was a subtle effect the crew almost certainly never noticed.

"Does this mean something to us? We're not sure," Dittemore told reporters. "It could be indicative of rough tile. It could be indicative perhaps of missing tile. We're not sure. We do know it's indicative there was an increase in drag on the left side of the vehicle.

"At this time, we also lost the left main landing gear tire pressure and wheel temperature measurements," he continued. "We're fairly confident that this loss of information was measurement related and not loss of the tires themselves, because the measurements were staggered in their loss. If we'd have lost a tire, we believe we would have lost all the measurements at the same time. That didn't occur."

At 8:59 a.m., as Columbia descended across west Texas, the flight computers ordered additional elevon adjustments to offset what appeared to be an increasing drag on the left wing. Whatever was wrong, it was getting worse, and fast.

"The flight control system was countering that drag by trying to command the vehicle to roll to the right-hand side," Dittemore said. "Soon after, we had loss of signal."

Dittemore told reporters that engineers believed it might be possible to extract data from an additional 32 seconds of telemetry beyond the point where Husband's final call to Mission Control was cut off, at 8:59:32 a.m. The shuttle was filmed breaking up a few seconds past 9 a.m., and engineers believed any data from that final half-minute of flight could be crucial for figuring out exactly what had happened.

But he didn't say what might have caused the unusual sensor readings in the left wing. Reporters were quick to latch on to the foam strike as a possible explanation.

"I know I'm thinking the same thing you're thinking, but I can't go beyond that," he responded. "I want to be careful that I don't jump to conclusions because if I do, I'll miss something else that may be very important. So you got the point. There may be some significance to the wheel well. We're going to look in that area more carefully."

Questions kept returning to the foam strike. Dittemore gave a brief summary of how the debris assessment worked during the flight and repeated the conclusion that the strike was inconsequential. "It was not going to represent an impact to our flight control qualities or our safety," Dittemore said.

"But even then, we talked about what if it did," he continued. "What would we do even if it did? We talked about, in our technical teams, is there a way that if we were wrong, was there a way during entry that would minimize [heating]? . . . "We asked ourselves, is there any other option? There's no other option. If you want to come back home, you have got to come back home through the atmosphere. We put it in such an attitude as to minimize the thermal effects on the vehicle."

Asked again about the decision not to seek close-up views of Columbia through military telescopes or satellites, he said NASA's past experience with such imagery indicated it wouldn't be conclusive or provide the sort of depth perception necessary to judge the severity of a strike.

"The second factor was, even if I had that information, I couldn't do anything about it," Dittemore continued. "I'm really helpless to go out and do any tile repair. And the third factor was I had done the analysis. The best experts at our disposal concluded that it was a minor problem, not a significant problem. And when you added all that up, there was no need to take pictures to document any evidence because we believed it was superficial and a turnaround issue . . . not a safety issue. And so, we didn't take any pictures."

At Barksdale, Gehman's accident board was swinging into action. For the first time, the board discussed the politically sensitive subject of the panel's independence from NASA, an issue that already was prompting some on Capitol Hill to question the objectivity of the investigation. The investigation's original charter, as stated in NASA's Contingency Action Plan, said the panel would:

> A. Conduct activities in accordance with direction from the NASA Administrator and the provisions of applicable NASA management instructions.
>
> B. Work with the NASA Administrator to schedule board activities, interim board reports and submission of the final report.
>
> C. Determine the facts, as well as the actual or probable causes of the shuttle mishap in terms of primary cause, contributing cause(s) and potential cause(s) and to recommend preventative and other appropriate actions to preclude recurrence of a similar mishap.

In parentheses, point C noted, "The investigation will not be conducted or used to determine questions of culpability, legal liability or disciplinary action."

In its Feb. 3 meeting, the board touched on issues that would, in short order, flare up into a political firestorm. According to the minutes of the meeting, the board agreed that to be independent "means having the board's own administrative staff (not rely on NASA support)" and "having the board's own technical experts (not rely on NASA support)."

"I knew that independence, autonomy from NASA, was going to be an issue," Gehman said later. "There was a draft charter that was faxed to me, right out of the manual, and right away there were some things in the draft charter that I knew I couldn't live with. It said do things that the NASA administrator says and clear all your activities through the NASA administrator. And there were some things in there that—[it] didn't take a rocket scientist to figure out that you couldn't conduct an independent investigation with provisions like that."

But in the first few days of the investigation, Gehman viewed those issues as relatively "minor administrative kinds of things."

"I don't want to give you the impression that I was clairvoyant here or something like that," he said.

But in a sense, he was. Those very issues were raising questions in Congress about whether Gehman's board could independently determine the root cause of the disaster or whether the nation would be better served by the appointment of a presidential or congressional commission.

Back at Johnson, Dittemore continued to hold daily briefings. He updated the accident timeline that afternoon, adjusting the sequence of previously discussed sensor failures, adding additional sensor dropouts in the left wing, and refining the initial temperature readings from other sensors that didn't immediately fail. Engineers remained baffled by the higher temperatures on the left side of the shuttle's fuselage above the wing.

There still was no direct evidence that Columbia's wing, landing gear wheel well, or fuselage had actually been breached. There were unusual, higher-than-normal temperatures in the wheel well and on the fuselage. There were multiple left wing sensor failures, and increasing aerodynamic drag on the left side of the shuttle, but no direct evidence of what might be causing those responses.

Dittemore also revealed that Columbia's flight computers had ordered two of four right-side rocket thrusters to fire just before loss of contact, to assist the elevons in the losing battle to keep the shuttle's nose pointed in the right direction.

"I mentioned yesterday that we had evidence of increasing drag on the left wing, that the aerosurfaces were reacting to that drag to maintain our attitude and trim," Dittemore said. "As we have continued to pore over the data, it's not the absolute value of the attitude change that is interesting. What is becoming interesting to us now is the rate of change."

In other words, the drag was accelerating so rapidly that it was forcing ever more extreme responses from Columbia's flight computers to keep the ship on course. The thruster firings began a few seconds after Husband's last call to Mission Control. Engineers were still working to extract data from the final 32 seconds, but it was proving more difficult than expected.

Dittemore faced more questions about the foam strike. He reviewed the assumptions that went into the debris analysis, including the foam's size, assumed weight and impact velocity. For the first time, he mentioned the Crater program, saying it overpredicted tile damage.

The *Orlando Sentinel* was working on a story for the next day that quoted anonymous sources who said they had been concerned about the accuracy of the debris analysis in the absence of photographs from spy satellites or military telescopes.

"Now I am aware, here two days later, that there have been some reservations expressed by certain individuals . . . so we're reviewing those reservations again as part of our database," Dittemore said. "They weren't part of our playbook at the time because they didn't surface; they didn't come forward. But now that the event happened and now that this data has come forward concerning some reservations or even their belief of the worst-case 'what-if' scenario situation, we're going to look at that and we're going to pay attention to it."

Finally, Dittemore was asked whether the foam strike was still on the NASA list of possible causes. His reply would play a major role in shaping news coverage of the disaster for the rest of the week.

"Certainly this debris is one of our primary areas of emphasis," he said. "We are completely redoing the analysis from scratch. We want to know if we made any erroneous assumptions. We want to know if we weren't conservative enough. We want to know if we made any mistakes.

"Secondarily, we have a team of engineers and managers and technicians that are working to understand the shedding of the debris itself from the external tank. And this is how we're approaching it:

We're making the assumption from the start that the external tank was the root cause of the problem that lost Columbia. That's our starting point when we look at the tank. And based on that assumption, what is the fault tree that would substantiate that particular assumption? And so we are attacking that.

"That's a fairly drastic assumption, and it's sobering. But we've asked our ET—our external tank—project management and their personnel and our contractor to make that assumption and see where that leads us. And so even though that's a drastic assumption, I think that's what we need to do."

Dittemore was describing how one team of engineers—the External Tank Project—had been asked to assume that the foam shedding had caused the disaster. Other teams were looking at other potential causes. But the comment was widely interpreted by the public and media as meaning NASA now viewed the foam strike as the leading theory for explaining the Columbia disaster. It would take another 24 hours for Dittemore to clarify his remarks. Along with virtually everyone else at Johnson, he took time off Tuesday, Feb. 4, to attend a memorial at the space center to honor Columbia's crew.

It was a glorious Texas morning. President Bush flew in, and hundreds turned out to hear his words of comfort to the astronauts' families, seated on folding chairs on a lawn in a scene reminiscent of another memorial 17 years earlier, when President Ronald Reagan addressed the Challenger families.

"Our whole nation was blessed to have such men and women serving in our space program," Bush said, his words broadcast live by television and radio. "Their loss is deeply felt, especially in this place, where so many of you called them friends. The people of NASA are being tested once again. In your grief, you are responding as your friends would have wished, with focus, professionalism and unbroken faith in the mission of this agency." Four T-38 jets flew overhead. One peeled off, forming a "missing man" formation in the cloudless blue sky.

Gehman and the board members had decided not to attend the astronaut memorial because "we thought that would be disruptive and distracting." Instead, they visited Nacogdoches, Texas, to meet with salvage workers and scheduled field trips to Kennedy, Marshall and Michoud for the following week. The goal was to get educated on the shuttle as quickly as possible.

"None of us knew a space shuttle from a shuffle board shuttle, so we immediately had to get real smart on the shuttle," Gehman said later. "As shuttle people started giving us briefs on not only how the shuttle worked but who was responsible for each part, every answer had the word 'JSC' [Johnson Space Center] in it. So we started saying, 'well, why are we sitting here on folding chairs at Barksdale Air Force Base?' "

The board flew to Houston Wednesday, Feb. 5, and set up shop. Working out of NASA conference rooms at first, they eventually rented space at an office complex a mile or so from Johnson.

Dittemore got his chance to correct Tuesday's headlines at a press conference that afternoon. This time, he swung too far the opposite way. He held up a large piece of external tank foam to demonstrate its light weight. He encouraged reporters to pick it up to "try to get an understanding of its consistency, which is not very hard; in fact, it's fragile, it's easy to break, and it's easy to break up into particles.

Dittemore again repeated the rationale engineers had used to dismiss the foam strike. "And so it's difficult for us to believe, as engineers, as management and as a team, that this particular piece of foam debris shedding from the tank represented a safety of flight issue.

"So we're looking somewhere else. Was there another event that escaped detection? As I mentioned before, we're trying to find the missing link. And as you focus your attention on the debris, we're focusing our attention on what we didn't see. We believe there's something else, and that's why we're doing a fault-tree analysis, and that's why we're investigating every area. Right now, it does not make sense to us that a piece of debris would be the root cause for the loss of Columbia and its crew. There's got to be another reason."

It was a mistake, causing another wild swing in the media's coverage and putting political heat on Dittemore from NASA headquarters.

"I came back after the memorial to try to balance [the foam question] out a little bit," Dittemore said later. "What I was trying to do was say, 'Look, it's not so clear to us that foam's the only thing; maybe there's a missing link that we haven't thought about yet.' The point was to put balance on the media story, because the media was off and running toward the foam, rightly or wrongly, and we didn't know if it was right or wrong."

At the time, O'Keefe was struggling to convince skeptics of the independence and credibility of the accident investigation. He went through the roof listening to Dittemore all but dismiss foam as a factor in the accident. Such decisions were up to the board, not NASA, and Dittemore quickly got the message.

"Certainly I got some feedback that he wasn't very happy that I'd said so much about it not being foam," Dittemore said later.

The next day, Thursday, Feb. 6, he attempted to set the record straight—again.

"As I talked to you yesterday, I mentioned to you that we believe in some instances that it's hard for us to understand why a block of foam that has fallen off the tank could have been the root cause," Dittemore said. "But that is not stopping us from continuing to investigate that particular event as being a potential root cause."

Earlier that day, O'Keefe and his senior managers, along with scores of other headquarters personnel and guests from Johnson and Kennedy, gathered at the National Cathedral in Washington for another memorial service. Vice President Dick Cheney spoke, Cabana told anecdotes about the fallen astronauts, and Patti LaBelle sang a stirring tribute originally commissioned to mark the 100th anniversary of the Wright brothers' first powered flight.

The accident board, meanwhile, spent Thursday listening to hours of NASA briefings and updates at Johnson. Whittle provided an update on the search and salvage operation. Flight director Paul Hill gave an overview of efforts to locate critical debris by analyzing possible tracks in air traffic control radar plots. He also updated the board on the progress of work to evaluate amateur camera views of Columbia's descent, saying he now had 15 high-resolution sightings. "Good images have been submitted that show Columbia was shedding debris as it crossed the California coast," the board minutes noted.

In other CAIB business, Gehman ruled out daily press briefings but agreed to hold periodic news conferences. "Periodic" eventually came to mean once a week. He met with Dittemore that day to discuss

NASA's remaining briefing schedule. The two men decided Dittemore would conduct a final press conference Friday and that Gehman would take over the following week. Dittemore's response: "Thank goodness. Yahoo!"

Friday's final briefing would include no surprises.

Earlier that morning, thousands of Kennedy workers gathered at the south end of the Shuttle Landing Facility runway to remember Columbia and its crew. Robert Crippen, pilot of Columbia for its maiden voyage in April 1981, provided a moving tribute to the lost spacecraft, a machine many at Kennedy had spent decades maintaining.

In the days and weeks ahead, Crippen's eulogy would be mentioned time after time by NASA and contractor workers at Kennedy as one of the most emotional moments in a week of memorials.

"It is fitting we are gathered here on the shuttle runway for this event," Crippen said. "It was here last Saturday that family and friends waited anxiously to celebrate with the crew their successful mission and safe return to Earth. It never happened. I'm sure that Columbia, which had traveled millions of miles and made that fiery re-entry 27 times before, struggled mightily in those last moments to bring her crew home safely once again. She wasn't successful.

"Columbia was a fine ship. She was named after Robert Gray's exploration ship, which sailed out of Boston Harbor in the 18th century. Columbia and the other orbiters were all named after great explorer ships, because that is their mission, to explore the unknown.

"Columbia was hardly a thing of beauty, except to those of us who loved and cared for her," Crippen said, fighting back tears. "She was often bad-mouthed for being a little heavy in the rear end. But many of us can relate to that. Many said she was old and past her prime.

"Still, she had only lived barely a quarter of her design life; in years, she was only 22. Columbia had a great many missions ahead of her. She along with the crew had her life snuffed out in her prime. . . . There's heavy grief in our hearts, which will diminish in time, but it will never go away, and we will never forget," Crippen finished.

"Hail Rick, Willie, KC, Mike, Laurel, Dave and Ilan. Hail Columbia."

NINE

ECHOES OF CHALLENGER

I truly believe in what Willie and our crew were accomplishing as explorers. But the reasons that the CAIB [Columbia Accident Investigation Board] report gave for the Columbia's tragic end were not "acceptable" risks. The vehicle should not have launched, and our loved ones did not have to die.

—*Lani McCool, wife of Columbia pilot Willie McCool*

The Columbia Accident Investigation Board made its long-awaited public debut inside a packed media auditorium at Johnson 10 days after the accident.

With Ron Dittemore happily stepping out of the limelight, the board—commonly known by its much-shorter acronym CAIB (pronounced KABE)—now was the public face of the investigation. The panel's first press conference, however, wasn't so much a progress report as a declaration of independence.

Eager to dispel concerns in Congress that the CAIB would be doing NASA's bidding, chairman Hal Gehman used the briefing to distance his investigation from the space agency. One of the first things the board had done, he said, was to create a technical analysis team independent of NASA.

"We have the beginnings of a team of independent experts who have nothing to do with NASA," Gehman said. "It's not that we don't believe the NASA data. It's just that in order that the report be truly independent and that the conclusions be based on good, solid analytical work, both inside NASA and outside NASA, this team will advise us on when we come across one of those junctures in the road."

To add to the appearance of separation, the board had moved its offices off NASA property at Johnson. They were setting up a website that would allow the public to contact investigators directly without going through NASA. Even future press conferences would be held away from Johnson's state-of-the-art broadcast facilities.

Nevertheless, reporters pounced on the independence issue when the time came for questions. Gehman was ready.

"The administrator of NASA signed the letter creating this board," Gehman said, "and his street address will be on the envelope of the outside of the report. But this board is aware that we have many constituents. We are fully aware that the families of the deceased astronauts are our constituents. We are aware that the Congress of the United States is one of our constituents. The White House, the taxpayers and the citizens of this country are all constituents of this board. . . . We are not the least bit concerned about how the various branches of government in Washington work out who is going to review whose work or anything like that, because ours is going to be a deep and thorough investigation. We're going to attempt to find the causes and make recommendations."

To those who knew retired four-star Adm. Harold W. Gehman, Jr., the thought of him as NASA's lapdog was laughable.

The board's 60-year-old chairman had grown up in Norfolk, Virginia—he still lived in the area—as the son of a Navy captain. After an unremarkable stint as an industrial engineering student at Pennsylvania State University, the Navy ROTC graduate went off to war in Vietnam. There, he commanded a small patrol boat that saw its share of action. Friends and relatives noticed a changed, more serious Hal Gehman when he returned. The Navy discovered he was as skilled as a manager as he was as a war fighter. His calm, self-assured manner projected authority and commanded respect—yet almost effortlessly so. He quickly moved up the ranks, commanding a missile cruiser, a destroyer and finally the Dwight D. Eisenhower aircraft carrier battle group. By 1996, he had been awarded a fourth star and assigned to the Navy's number two post. As vice chief of operations, he was in charge of more than a third of a million people and a $70 billion budget, almost five times larger than NASA's. He was named NATO's Supreme Allied Commander, Atlantic, and head of the U.S. Joint Forces Command in 1998. He walked away from it all in 2000 to spend more time with his wife, two grown children, siblings and parents.

The attack on the USS Cole happened shortly after Gehman's retirement. He and a retired Army general were chosen to head the investigation. The result was a tough, candid assessment that uncovered major problems, including overconfidence, in the military chain of command. The report minced no words in attacking widely held misconceptions about the threat posed by terrorists. NASA administrator Sean O'Keefe was familiar with the Cole Investigation from his days as a professor at Syracuse University, when colleagues used the probe as a case study in a professional development program. Gehman's investigation had explored why the attack succeeded. Parallel investigations separately were looking at other issues, such as who was to blame. It quickly became clear to Gehman that if the CAIB report was to have any real impact, it would have to go beyond the technical causes of Columbia's breakup to explore deeper human and institutional failures.

As Gehman faced reporters for the first time, he knew the board's credibility was at stake. He understood something had to be done to end the concerns about independence once and for all. Before the investigation was a week old, he had fought for and won major changes in the board's charter that went a long way toward accomplishing that. The CAIB would have an independent budget. It would have full control over the final report. References in the charter to the board working "with direction from the NASA administrator" were eliminated. The board's mission would be expanded to finding "the actual or probable causes of the shuttle mishap in terms of dominant and contributing root causes and significant observations." And new board members, civilians not specified in the original contingency plan, would be added.

Two new hires quickly were brought aboard. Roger Tetrault, a retired former executive at energy company McDermott International and defense contractor General Dynamics, was hired the week after the accident. Sheila Widnall, an engineering professor at the Massachusetts Institute of Technology and former secretary of the Air Force, joined a week later. Three final members were named March 5: Sally Ride, a physics professor at the University of California, San Diego, the first U.S. woman in space, and a member of the Rogers Commission that investigated the Challenger accident; Doug Osheroff, a physics professor at Stanford University and 1996 Nobel Prize winner; and John Logsdon, director of the Space Policy Institute at George Washington University.

Gehman was fond of saying that any of the CAIB's members had the credentials and mental horsepower to chair it. There was truth to the statement, but it was Gehman's investigation and no one doubted for a moment who the boss was. His tone, however, was collegial and his manner was contagious. One of the board's 120 staffers remarked that Gehman was someone who showed the same courtesy and respect to the person who emptied the trash as he did to fellow military officers. Other board members noticed.

"It was a very interesting collection of cultures," Scott Hubbard said. "I really give Admiral Gehman a lot of credit for being able to bridge across so many backgrounds. . . . He did, I think, an outstanding job of forging a consensus out of this."

The board quickly began to gel as a team. There were a few personality conflicts, but for the most part, the eclectic members genuinely came to like each other during the seven-month investigation. Early on in Houston, before moving to Washington to write the final report, they lived together in an extended-stay hotel near their Regents Park offices, worked together during the day, then relaxed together after hours at spots like Molly's Pub, a not-so-upscale watering hole in Clear Lake.

One evening in late May, the board had just finished outlining the report and assigning who would write what sections. A celebratory trip to Molly's was in order.

"After a couple of beers, we decided that since we had finished the outlining of the report—we considered this essentially the script—that now we should cast the parts of 'CAIB: The Movie,' " Gehman said. "So somebody got a napkin out and started taking down all of the key people and deciding who was going to play what roles. I remember laughing so hard tears came to my eyes. It probably was the beer talking."

The slim, distinguished Gehman would be portrayed by veteran leading man Kirk Douglas. Shirley MacLaine was penciled in to play Widnall. Dustin Hoffman would get the role of Osheroff. There were nicknames also. Hubbard, for example, became known as "Hollywood" because he seemed to enjoy the limelight more than others. Jim Hallock, a mild-mannered researcher, was dubbed "ZZ Top" because of his resemblance to the rock group's bearded musicians.

The esprit de corps fostered by Gehman helped make easier the long hours of work, including the grueling road trips by board members and investigators to NASA facilities across the country. While some of the board's academics tried to balance the CAIB work with their regular duties, most of the panel, including the military officers, took the investigation on as full-time jobs.

The collegial atmosphere, however, didn't prevent frequent squabbling and debate about the investigation and the report's final form. Sometimes, those debates became heated.

"I forget if it was Steve Wallace talking to me or me talking to Steve Wallace, but one of us gave the other an honorary OBE, not Order of the British Empire, but Opinionated Blowhard, Esquire," Hubbard remembered. "This was serious business. We wanted to find out why we lost an orbiter and seven people died—and there were lots of strong personalities."

To investigate the accident, Gehman divided the board into three subgroups. Group 1—John Barry, Duane Deal and Steve Turcotte—would investigate shuttle maintenance, materials and management issues. Group 2—Ken Hess, Ride and Wallace—would look at operations, broader organizational issues, training, and flight readiness.

Group 3—Hallock, Hubbard, Osheroff, Tetrault and Widnall—would look at the technical cause of Columbia's breakup. A fourth group—Logsdon, with help from Gehman and Ride—was added later to explore cultural, historical and policy issues like the shuttle's origin and NASA's budget.

The first of the board's 10 public hearings took place March 6 in a theater on the University of Houston's Clear Lake campus. Ron Dittemore was among the witnesses. Those expecting fireworks, however, were disappointed. His testimony was mostly limited to generic questions about NASA's chain of command, managers' roles and the relationship between the space agency and its contractors. There was no mention of what happened during Columbia's flight and Dittemore's role in critical decisions.

To the disappointment of reporters, the nine subsequent hearings would take the same approach. The public meetings would be used to gather background information and discuss big-picture issues of concern to the CAIB. There would be no public grilling of shuttle managers or engineers on their actions. Instead, that would be done behind closed doors under the promise of privilege.

The controversial decision to grant witnesses confidentiality had been reached within the first two days after the accident. Bryan O'Connor, briefly a non-voting CAIB member, already had urged that interviews be kept secret and not be sworn testimony but rather "witness statements." The CAIB's military representatives agreed and argued that investigators would learn more if witnesses knew their remarks wouldn't be made public. Congress was pressuring Gehman to dig deeper than the Rogers Commission that investigated Challenger to find any larger institutional problems. Gehman decided privilege would make that job easier.

To accomplish this, the CAIB took advantage of a law that allowed boards made up entirely of federal employees to meet and conduct business in secret. The seven original CAIB members all fit the requirement as either military personnel or employees of federal agencies. Retiree Gehman was hired on by the federal Office of Personnel Management. The five board members added later were put on the NASA payroll, a move that again raised the issue of independence.

They vehemently denied, however, that who signed their paychecks would affect the investigation. The proof, they said, would be in their final report.

"I don't see it as an issue for the board members to be on the federal payroll—this board, unlike most pro-bono government committees, is essentially a full-time job (for which people should receive some compensation)," Ride explained in an e-mail to the *Orlando Sentinel*. "But one might ask whether it should be NASA's payroll. . . . Since the White House hasn't picked up the mantle on this investigation, but rather has left it to NASA, I don't see an alternative payroll source—or alternative source for funding the investigation itself."

Initially, Gehman and the board also wanted to keep the interviews secret from Capitol Hill because of congressional staffers with close ties to NASA and fear of leaks to the press. The ban became a thorny issue that finally was resolved after the CAIB agreed to give lawmakers and staff limited access to interview transcripts in a closely monitored reading room.

The board ultimately would interview 210 people under privilege. When the investigation ended, opinions were split on how much the board learned by granting witnesses confidentiality. Some CAIB members felt strongly that privilege had proven an important tool in learning how NASA operated. Others, like Gehman, weren't so sure.

"The interviews, by and large, were not that spectacularly useful." Gehman said. "I think that 99 percent of everything that any one of those people said they would be willing to say in public. So they weren't a great gold mine. They did, however, [say] some pretty straightforward things about safety and their bosses. . . . They did confirm for us in the strongest possible sense that when we started this investigation into the organizational causes that we were on the right track. I mean, we have not 1, not 2, but 15 or 20 people who corroborated each and every point."

Perhaps the biggest benefit of gathering privileged testimony, Gehman added, was that it gave investigators the ability to cover a lot more ground than they could have otherwise.

"Sometimes we were doing seven or eight interviews a day," Gehman said. "If you tried to do that in a public hearing, it just would have taken forever. I don't believe we would have gotten the quality job that we did."

The board had its work cut out. As the investigation got off the ground during the first two weeks after the accident, it wasn't yet clear how closely they would have to scrutinize shuttle management. In the days immediately after the disaster, the board's main focus was on a simple question: What happened? By Feb. 13, the board already had concluded "the temperature indications seen in Columbia's left wheel well during entry would require the presence of plasma" and "heat transfer through the structure as from a missing tile would not be sufficient to cause the temperature indications seen in the last minutes of flight." Engineers' fears during the mission of a damaged seal on Columbia's landing gear door looked like a very real possibility. No one, however, could make any scenario fit the baffling assortment of sensor readings and other data transmitted down during the doomed orbiter's return.

Slowly but surely, investigators would find some of the answers in the woods and fields of east-central Texas, as the largest recovery operation in U.S. history gathered steam.

Carl Vita and Marty Pontecorvo still shake their heads in disbelief.

It was one week after the Columbia disaster. They had been called to a home near Palestine, Texas, to inspect what a resident believed was a piece of shuttle debris. Vita, a veteran engineer with United Space Alliance at Kennedy, pulled his pickup truck off the two-lane country road and parked. The residents weren't home, but they had placed a small American flag by a foot-long piece of aluminum wreckage. Vita and Pontecorvo, a senior NASA engineer at Kennedy, marked the location of the debris and looked around nearby for anything the homeowner might have missed. It was all standard procedure.

As they walked back to their truck, Vita noticed what looked like a discarded audio cassette in the dirt by the road. He must have driven over it while pulling over to park, the truck's wheels passing to either side. The tape was partially unwound.

"I walked around the back of the truck and saw the tape sitting there, just down on the ground. It was about 3 inches off the side of the pavement in the dirt," Vita recalled. "I kind of disregarded it at first. It was like when you see somebody throw a cassette out their window. I'm thinking audio tape, even maybe an eight-track."

The tape didn't look quite right, though. It was wider than a normal audio cassette, and the plastic case had only one visible reel. Vita decided to pick it up anyway. He told Pontecorvo, "It'll probably turn out to be Waylon Jennings or a Merle Haggard tape, and the lab guys will get a kick out of it."

All search crews were told to store magnetic tapes in special bags that prevented a buildup of static charges that might corrupt whatever data were present. Pontecorvo realized they didn't have any of the special bags, but quickly came up with a suitable alternative.

"Wal-Mart had bought us lunch that day, fried chicken. They gave everyone a couple of pieces of chicken and a biscuit and a little thing of cole slaw in a little bag," he said. "I had my garbage in it. I dumped the garbage out of the bag into the car and put the tape in my greasy fried chicken bag. And that's how it went to Johnson."

The two didn't give the find a second thought. Like scores of other salvage teams spread out across central Texas, Vita and Pontecorvo were too busy to waste time pondering a recovery that probably wouldn't amount to much. As shuttle experts, their job was to rush to sites where debris had been spotted, to make a positive identification and if the debris was from Columbia, to arrange for its recovery. Texas police officers got so used to seeing the NASA rental cars flying past "they'd wave at you," Pontecorvo laughed. "They didn't even pull us over anymore. It was insane out there."

Not all of the calls turned out to involve shuttle debris.

"We went to this one place, and the lady had bird poop on the side of her house," Pontecorvo said. "And she says, 'I've got shuttle fuel on my house.' They told us when we went out there, they said just do whatever they ask. So we hosed this bird poop off the lady's house. I said, 'How do you know it's from the shuttle, ma'am?' She said, 'It wasn't there before Saturday.' I said, 'Oh, OK.' "

Another man called about a 4-foot-wide hole burned in his wooden dock. Divers were called in "and they found one of those big fire pots, those barbecue things. His grandkids were over there and got it too hot, got it on fire, and it burned through the deck."

Most of the work, however, involved the grim recovery of shuttle hardware, crew equipment and, in the first few days, human remains. While the center of activity was in Texas and Louisiana, the CAIB later noted that NASA received reports of debris from 37 states that Columbia didn't cross, as well as Canada, Jamaica and the Bahamas. Searchers focused their efforts instead on the seven states the shuttle did fly over, where pieces were believed to have fallen. Police officers checked a stretch of the California coast for wreckage that might have washed up. Prison inmates scavenged for parts of the shuttle in the Nevada desert. Volunteers combed parts of Utah, Arizona, and New Mexico. But in the end, search crews found debris from Columbia in only two states: Texas and Louisiana.

After the first two weeks, more than 3,000 people were at work on the ground, and by the end of the first month, more than 4,000 had been flown in from around the country. Search areas were divided into grids, and 20-person teams were dispatched to walk 10 feet apart in lines, fanning out across fields, swamps and woodlands. The teams covered an area 2 miles to each side of Columbia's ground track. According to the CAIB's final report, ground crews were finding more than 1,000 shuttle pieces daily at the peak of operations.

Once debris was recovered, it was photographed and its location recorded. Four Texas collection centers were set up at Corsicana, Hemphill, Nacogdoches and Palestine. From there, pieces were shipped to Barksdale Air Force Base, and then onto the debris hangar at Kennedy, where technicians performed a final identification and positioned the items on a giant reconstruction grid.

In addition to the ground search, dive teams used sonar to map the bottom of the Toledo Bend Reservoir and Lake Nacogdoches in Texas. Tons of debris had been recovered on the ground in surrounding areas, and sonar detected almost 3,500 objects at the bottom of the two bodies of water. Divers, however, were unable to see more than a few inches in the murky water and recovered only one piece of Columbia hardware.

More than 40 helicopters and aircraft were used in the search to survey areas up to five miles away from Columbia's ground track with high-tech scanners and radars. Tragically, a Forest Service helicopter had mechanical trouble and crashed on March 27 near Broadus,

Texas, killing pilot Jules "Buzz" Mier and Texas Forest Service employee Charles Krenek. Two Kennedy workers and a Forest Service employee from South Dakota were injured.

For the most part, residents were conscientious about turning in debris. Several items, however, began to turn up on Internet auction sites. NASA made it clear that all wreckage was government property and anyone keeping pieces as souvenirs could be prosecuted. The space agency offered a brief amnesty period that allowed people to turn over any finds without fear of legal action. Some didn't heed the warnings. Several, including a quality assurance worker at Kennedy, eventually were arrested.

Vita and Pontecorvo turned in the cassette they found and didn't think about it again until they got back to Florida at the end of February.

By then, technicians at Johnson had analyzed the tape and discovered what it was. The weekend Vita and Pontecorvo got home, it was playing on national television. They weren't especially surprised to find out it wasn't a country music tape. Their jaws dropped however, when they discovered the battered cassette they had found in the dirt beside a country road in Texas contained 13 minutes of videotape shot by astronaut Laurel Clark during Columbia's re-entry.

"It was pretty bizarre," Vita recalled. "Where it was at, right on the side of the road, it was just 'Ah, come on, somebody threw this out the window as junk.' "

But it wasn't junk. It was a final glimpse of the four astronauts on Columbia's flight deck—Husband, McCool, Chawla and Clark—all of them in good spirits and looking forward to landing in Florida. The outer portions of the tape had been burned away during the camera's fall to Earth, and the video ended more than a minute before the first signs of trouble. Many saw that as a blessing for the families, who approved the tape's public release.

"They were still living large and having a great time," astronaut Jerry Ross said of his fallen colleagues.

As investigators began piecing together the debris puzzle collected by salvage crews, other CAIB members were setting their sights on NASA's safety organization and management.

Details of the unsuccessful photo requests, concerns about the debris analysis and e-mail traffic between worried engineers already were surfacing in the media the week after the accident. Board members later would say there was no single "Eureka!" moment when the CAIB realized NASA had big management problems. But the more the CAIB learned, the clearer it became that the probe would have to delve deeply into what investigators eventually would call "a broken safety culture."

The board was baffled by how NASA could have missed the danger posed by the foam when its own hazard studies showed the material was capable of causing a shuttle catastrophe and killing the crew. What emerged was a portrait of isolated shuttle safety organizations at different NASA centers that weren't coordinated across the agency. Those organizations, investigators found, were often poorly managed and depended on the programs they were supposed to monitor for resources.

Several months later, Gehman would sum up the problem during a May appearance before the Senate Commerce Committee: "We find the safety organization on paper is perfect, but when you bore down a little deeper, you don't find any 'there' there."

Gehman and other board members had been floored during a March 25 public hearing at Cape Canaveral by the testimony of Bill Higgins, Kennedy's safety chief for shuttle processing. The longer Higgins spoke, the more confusing it became as to whom his office answered to.

"I hate to be dense about this," Gehman told Higgins. "You've attempted to tell me this twice already. Maybe the third time it will work. I'm still confused about who you work for and what your organization does. . . . You said if [Kennedy director Roy] Bridges wanted to give you some more people, you would know how to put them to work. . . . I thought this was [a] shuttle program [function] and shuttle-funded, in which case you should have said 'If Mr. Dittemore wants to give me some more people.' Or have I got it wrong?"

"Well, you've got me on this one," Higgins replied.

Later in the hearing, Gehman asked Higgins where he would find the safety organization that looked at big-picture concerns like falling foam as opposed to those checking daily processing to make sure bolts were properly tightened in the shuttle's hangars.

"Where is that organization, and where is that place in the food chain that we should be looking at?" Gehman said.

"That's a very good question," Higgins replied. "I'm not sure I can answer that specifically because that would be program and agency functions that are above me."

Gehman later would call Higgins the single most important person the CAIB questioned during one of its public hearings.

"He testified that he didn't know who he worked for," Gehman explained. "Boy, that set off alarm bells. . . . He couldn't remember who he worked for and contradicted himself. The board wrote that down—'The safety guy doesn't know who he works for.' "

Further investigation made it obvious the disconnect wasn't just at Kennedy. NASA headquarters' safety office had failed miserably to coordinate functions with all of the agency's field centers.

Gehman had CAIB staff members sit in on NASA technical meetings to observe the interaction between engineers and managers. People with safety concerns, like Rodney Rocha, had failed to appeal decisions such as the photo request to top-level program officials. The board wanted to know why. Gehman later recalled what investigators found:

> These [staff] people started telling us stories of not very positive management techniques. [If] somebody had a different opinion than the party line, then they were intimidated or ridiculed or things like that. . . . There is not one organizational trait that we complained about that we didn't witness for ourselves. We were actually in the room or we were there and saw a safety person do nothing, or we saw engineers being ignored and things like that. We actually saw it for ourselves. Slowly, over a period of many, many weeks, we came to the realization that there was more here than faulty foam. We started asking ourselves "How do we study and come to conclusions and base those conclusions on just as much expertise in this social engineering area as we are in physics and thermal dynamics in the physical cause area?" We said, "Well, we have to base it on a whole bunch of experts, just like we have experts in the other areas." So we started searching for who those experts were.

It didn't take long for the board to find just the right expert: Diane Vaughan, a sociology professor at Boston College. Vaughan, an expert on organizational analysis, authored a 1996 book titled *The Challenger Launch Decision: Risky Technology, Culture, and Deviance at NASA.* The book examined the Challenger accident in great detail, citing an "incremental descent into poor judgment, supported by a culture of high risk technology," as major underlying causes of the earlier disaster. In that case, NASA managers accepted problems with booster seals as normal, or in family, despite repeatedly being presented with evidence to the contrary—a process Vaughan referred to as the "normalization of deviance."

The parallels were unmistakable. Already, the CAIB had noticed growing similarities with the Challenger accident: increased acceptance of risk; concerns of engineers that were ignored or never heard; schedule pressures; poor communications. The realization that Columbia's breakup was rooted in the same sorts of failures had been articulated at an April 8 press conference by Ride, who had the added authority of having served on the Rogers Commission.

"I think I'm hearing an echo here," Ride observed, in one of the CAIB's most memorable lines. "During the Rogers Commission, one of the things that came out early on was that the O-rings were not a problem for the first time on this flight. . . . They had been a problem on not just one, not just two, not just three, but several shuttle flights before the Challenger accident. . . . [There was] the famous discussion of [Rogers Commission member] Richard Feynman that you survived it the first time, so suddenly it becomes more normal and it happens enough and now it's a normal occurrence. We're trying to understand whether that same thinking crept in with the foam off the tank."

Vaughan testified before the CAIB at a public hearing in Houston on April 23. Like a professor lecturing a seminar of eager graduate students, Vaughan methodically laid out the case for how NASA and its safety culture failed similarly during Challenger and Columbia's final flights.

Logsdon noted that Vaughan's book had been used to help improve a number of safety programs, including one for nuclear submarines. He asked her if she also was contacted by NASA after the book's publication.

"No," Vaughan replied, "though, in fact, as you said, the book did get quite a lot of publicity. I heard from many organizations that were concerned with reducing risk and reducing error and mistakes. The U.S. Forest Service called, and I spoke to hotshots and smoke-jumpers. I went to a conference the physicians held, looking at errors in hospitals. I was called by people working in nuclear regulatory operations, regular businesses, where it wasn't risky in the sense that human lives were at cost. Everybody called. My high school boyfriend called. But NASA never called."

The room erupted in laughter, but privately many NASA managers were seething. Some rejected the Challenger comparison out of hand, although copies of Vaughan's book increasingly could be found on desks at Johnson, Kennedy, Marshall and NASA Headquarters. A typical reaction was that expressed by chief flight director Milt Heflin the following day in an e-mail Vaughan still had months later. It was titled "sound bite . . ."

His message read: " 'My high school boyfriend called, but NASA never called' . . . a very cheap shot. . . . You kind of had me interested until then . . . too bad."

Nevertheless, Vaughan's message struck a chord with others at NASA as well as board members. Shuttle manager Trish Petete remembered hearing Challenger comparisons around Johnson as pressure mounted to keep the space station project on schedule.

"I was not part of Challenger," Petete said. "But other folks who had been were whispering in my ear saying 'This is Challenger all over again.' And I would say 'I hope not.' "

Eventually, Vaughan was hired by the board as a consultant. She would author a chapter in the CAIB's final report that compared the two shuttle disasters in an effort to show organizational changes needed to prevent a third. The shuttle's Mission Management Team, in particular, was one of those organizations where the board found changes were essential.

The MMT was created to oversee space flights from launch to landing. NASA rules stated the group was to meet daily during missions, although the requirement had been ignored for several flights. During Columbia's mission, the MMT met five times over a 16-day period. When it did convene, there was little significant debate. Managers focused on the bottom line instead of delving deeper into issues and questioning conclusions. Lower-level managers and engineers often were reluctant to speak up. The meetings seemed almost rehearsed, and the outcome of many key decisions was known in advance.

"The MMT is largely an official way of recording the results of activities that everybody already knows have been accomplished," said Dan Diggins, an FAA investigator who worked for the CAIB. "So to characterize the MMT as people who already have made up their mind, yeah, that's correct. But that doesn't mean they weren't open on the activity that went into it. It's just a pro-forma way of making a record."

Loren Shriver, a former shuttle commander and deputy program manager for United Space Alliance, said the MMT wasn't always like that. After three years of not regularly attending meetings in the late-1990s, he noticed a major difference when his job required that he resume going in 2000. The group no longer was a deliberative body.

"I said, 'Wow, this meeting has really changed a lot,' " Shriver recalled. "It was more of a meeting that heard primarily very top-level summary briefs on any event that happened and relied a whole bunch on off-line meetings and analysis to conduct business. [It] was primarily a very short crisp report out, and that was about it."

Concerns like the debris assessment team's request for a photo all too often hit a brick wall before ever reaching top program managers like Dittemore and Linda Ham in an MMT meeting. Rather than jump rank and defy protocol, some engineers, like Rocha, would give up or convince themselves everything was okay. Rocha characterized the flow of information as typically going from the top down, with managers above not always interested in having data flow up.

Other NASA managers hotly disputed such claims. But investigators increasingly adopted the view that the photo request was killed, in part, because it originated among mid-level engineers like Bob Page and Rocha.

"If the crew on orbit calls down and says 'Houston, we've got a problem' . . . [the shuttle program] comes forward to hammer that problem out completely until it's resolved to everyone's satisfaction," Diggins said. "This was different. This was not identified at Mission

Control. This was not identified by the crew. This was identified by some photo guy at Kennedy."

Some investigators like Diggins simply didn't believe the claim by Dittemore, Ham and others that they never knew the debris assessment team had requested photos.

"This debris team wanted imaging and that message got to Langley, it got to headquarters," Diggins said. "People knew this team was interested in getting imaging at other places in NASA. It's hard to believe that two buildings over that they didn't know about it. I just find that incredible. It's possible. But I think there might be some other explanation. Can I prove it? No. Either way it's a bad answer. If they didn't know, that goes right to the point of how bad their management was. If they did know and they chose to ignore it, that goes right to the point of how bad their management was."

Diggins added that the e-mail traffic only went so far in uncovering the truth.

"They know exactly what we [investigators] have in writing," he said. "Because of that, they are only going to admit to what it is they think we know. That's my view. I think they are wonderful people. They are very intelligent people. But they are in a defensive posture . . . Did we ever get to the bottom of this? No."

As Diggins and other investigators dug deeper, they found communication problems were merely the tip of the iceberg. NASA's culture gradually had evolved to the mindset where engineers had to prove conditions were unsafe, rather than safe to fly—a clear 'echo of Challenger' that went unheard. The shuttle program also was using the fleet like operational aircraft instead of the experimental ships they truly were. An example was NASA's habit of waiving requirements for crucial shuttle items.

The program's Critical Items List tracks almost 5,400 potential hazards in an effort to manage and control the risk. Some 4,222 are listed as "Criticality 1/1R." The failure of a "crit 1" item would result in the loss of the shuttle and its crew. The failure of a Criticality 1R item would leave the shuttle one malfunction away from a catastrophe. The program can issue a waiver when an item doesn't meet its specifications but the problem isn't deemed enough of a threat to replace the component. At the time of Columbia's launch, 3,233 of the fleet's Criticality 1/1R items had some sort of waiver. A third of those waivers had not been reviewed in more than a decade.

Equally baffling to investigators was the shuttle program's treatment of in-flight anomalies. The determination of whether a problem was dubbed an IFA often seemed arbitrary and based more on the whim of those attending a given Program Requirements Control Board meeting than clearly defined standards. The foam was a case in point. Every known example of bipod ramp foam loss had been designated an IFA until Atlantis' launch the previous October. Following that flight, however, shuttle managers assigned an action to the External Tank Project instead of making a formal classification. The action then was deferred until after Columbia's launch, meaning the issue wasn't even discussed at that mission's Flight Readiness Review.

It was unclear how many IFAs eventually were fully understood and fixed. Gehman observed during a CAIB meeting on March 28 that some of the issues "just seem to go away."

Marshall's External Tank Project had seen no need for an IFA on the foam loss, despite the fact an engineer postponed signing a preflight readiness statement until concerns about foam impacts were addressed. Investigators found that prelaunch assessments like the tank project's put managers in the position of overseeing a review of their own activities—another example of a flawed safety culture. Those assessments then went forward to the larger Flight Readiness Review, where they often were accepted at face value.

While much of the board's focus was on NASA, the shuttle program's contractors didn't escape scrutiny either. In 1996, NASA had begun consolidating dozens of shuttle contracts under one prime contractor, the newly formed United Space Alliance partnership between Boeing and Lockheed Martin. The resulting six-year Space Flight Operations Contract was billed as something that would save NASA more than a half-billion dollars annually by the start of the next decade. With NASA's blessing, contractors began moving civil service jobs to their own payrolls or eliminating positions entirely. Although the promised savings never materialized, the effort continued the large-scale downsizing that had begun five years earlier. All totaled, the shuttle's overall workforce dropped during the decade preceding Columbia's launch from 30,091 to 17,462—a 42 percent reduction.

There were plenty of warning signs that something was amiss. The short circuit that knocked out a critical computer during Columbia's July 1999 launch sparked a review by an independent assessment team led by then-Ames director Harry McDonald. The team's scathing report found, among other things, that "The workforce has achieved a conflicting message due to the emphasis on achieving cost and staff reductions, and the pressures placed on increasing scheduled flights as a result of the space station." The report also documented dozens of close calls where serious problems escaped detection that "could have caused a failure, luck or providence prevented it." Like dozens of other shuttle studies, the McDonald Report had little long-term impact. Many of its themes, however, would be echoed by the CAIB. Gehman called the report "brilliant."

As NASA jobs were transferred or eliminated, many safety functions formerly performed by civil servants shifted to contractors. Gradually, NASA began losing technical expertise and became increasingly reliant on contract workers. Tens of thousands of Government Mandated Inspection Points for the shuttle were eliminated. Wallace told CAIB members during a March 27 internal briefing that his group was "starting to see a picture of top-level meetings that are run by contractor senior executives with a NASA GS-13 [mid-level] safety representative sitting in the third row," according to minutes of the meeting. "This phenomenon is repeating itself."

Investigators, as well as some NASA officials, suggested decisions influenced by the bottom line had led to moves that were not in the best interests of safety. An often-cited example was the transfer of some Boeing shuttle operations, including analyses performed with the Crater computer model, from the Huntington Beach office to Houston—a transfer approved by NASA.

Astronaut Ross' view was widely held among the NASA rank and file: "When Boeing moves a lot of their engineering support here [Houston], guess what? The seasoned, experienced smart people wanted to stay back in California where they had their roots. So what did they fill in with? A bunch of young dudes who didn't have any of the history, didn't know the pedigree of the program, didn't know any of the history of the hardware prior to the time they signed on."

Others at NASA shared Ross' frustration. In the months following the accident, many had grown increasingly angry that while the CAIB put a relatively small number of individual civil servants under a microscope, vast numbers of contractor managers who now ran the program's daily operations had escaped scrutiny almost entirely. Some on the government side speculated that investigators were afraid to be too critical of industry because of unspecified legal issues.

By 2002, the makeup of the shuttle workforce had shifted to 15,744 contractors and 1,718 civil servants. To NASA officials like Dittemore, that meant there was more than enough blame to go around.

"That skill base migrated over to the contractor," Dittemore observed months later. "That included not only the bottom-level engineering, but also mid-level management and organization-level management of these engineering organizations. . . . NASA had to rely on the contractor input, involvement and their management positions. Certainly, we would challenge their decisions, but we had to rely upon their involvement. I find it interesting that as we go through this process of the Flight Readiness Review and the Mission Management Team, that there is very little contractor management conversation or involvement I could see. . . . I don't have any management on the contractor side coming forward and telling me things aren't right. I have only silence. . . . When a chunk of foam comes off, it's the contractors' job to analyze that and tell whether it represents a problem. The contractor has the skill base. When the contractor comes forward and says we've done our job and we don't believe there is an issue, they become part of the system and they are accountable for that recommendation. . . . And so when the board is critical of conversations that did or did not come forward to program management on the government side, I wonder where the contractor management was."

Mike McCulley, a former shuttle pilot and president of United Space Alliance, said Dittemore was simply wrong.

"I disagree completely," McCulley said. "We were not silent. We brought issues to the table, our management team supported our technicians and inspectors when they found things that resulted in the grounding of the fleet. . . . We make heroes of folks who find things that ground the fleet. It's our management that does that."

• • •

As the investigation gathered momentum, relations between the CAIB and NASA inevitably began to cool. Publicly, O'Keefe had unconditionally backed Gehman's request for changes in the board's charter. Privately, the NASA chief was said to have grown increasingly exasperated with Gehman's independent streak.

O'Keefe was the quintessential Republican Party loyalist, a talented career bureaucrat in the best sense of the word. The affable, 47-year-old had been in public service for virtually his entire adult life since graduating from Loyola University in his hometown of New Orleans. After earning a master's degree in public administration from Syracuse University in 1978, he was selected as a presidential management intern. That led to an eight-year stint with the Senate Appropriations Committee, where O'Keefe became staff director on the defense appropriations subcommittee.

When the first George Bush became president in 1989, Defense Secretary Dick Cheney brought O'Keefe to the Pentagon as comptroller and chief financial officer. His performance earned him a brief assignment in 1992 as Secretary of the Navy during the Bush administration's last six months in office. Like other Republican executives, he found refuge in academia for the Clinton White House's two terms, teaching at Gehman's alma mater, Pennsylvania State University, before returning to Syracuse. His exile ended in 2001, when George W. Bush and Vice President Cheney took office. O'Keefe was named the number two person in the White House's Office of Management and Budget, where he played an integral role in shaping the administration's fiscal policies.

One of the first issues O'Keefe tackled was a projected overrun on NASA's space station project estimated to reach almost $5 billion over a five-year period. In fact, the space agency's bookkeeping was so bad it was impossible to tell the true extent of the shortfall. O'Keefe helped draft a 2001 budget blueprint for federal agencies that solved the problem by slashing two modules from the station. When Bush tapped him in December to take over at NASA, the announcement sent shivers down the spines of the agency's true believers. It was obvious O'Keefe wasn't coming to stake out bold new initiatives for exploring the universe, a perception the new administrator quickly reinforced by declaring he was "a budgeteer, not a rocketeer." Instead, he was coming to put NASA's mishandled financial house in order. The station cuts earned him a nickname: "O'Grief."

After the Columbia accident, the last thing the Bush administration needed, on top of a war in Iraq and a slumping economy, was a lengthy shuttle investigation that eviscerated NASA and recommended costly, sweeping changes. So when Gehman began to stake out new territory for the CAIB, O'Keefe and Gehman's cordial relationship turned increasingly tense. That tension, however, wasn't all bad from NASA's perspective. It added to the growing perception that the board was, in fact, independent.

The situation came to a head the last week of February. The board had been troubled from the start that Ham and Ralph Roe, two of the most visible members of Columbia's Mission Management Team, continued to play pivotal roles in the investigation. Ham was leading NASA's Mishap Response Team, and Roe was the de facto director of the accident's engineering efforts. Investigators saw a looming conflict with the pair helping to guide an investigation that ultimately would examine their own performance. Gehman sent a letter to O'Keefe on Feb. 25 requesting that NASA "reassign the top-level space shuttle program management personnel who were involved in the preparation and operation of the flight of STS-107 back to their duties and remove them from directly managing or supporting the investigation."

O'Keefe had a videoconference with Ham and Roe. He told the managers that Gehman wanted them to quit the investigation. Neither was happy.

"O'Keefe felt that we could continue in our positions, and he was willing to defend us," Ham recalled. "But he wasn't sure if Gehman was going to follow through."

Gehman did. The following day in the CAIB's Houston offices, he met with Ham and Roe to tell both they needed to step down from the probe.

"I asked him, 'What are the reasons?' " Ham recalled. "He said, 'You can't investigate yourself.' Really, in hindsight, you know, he was right. You can't."

O'Keefe initially refused Gehman's request. In a Feb. 28 letter, O'Keefe wrote, "I am convinced this course of action will be viewed as prejudging the facts before the investigation is complete. Despite your assurances that no conclusions have been made arising from the facts of the investigation at this time, simply reassigning personnel will not accomplish your stated goal."

The following day, Gehman posted both letters on the CAIB's Internet site. O'Keefe got the message. Ham and Roe were officially taken off the investigation, although Roe continued to play a major role behind the scenes. Once considered the heir apparent to Dittemore as shuttle program manager, Roe was indispensable to the effort as one of NASA's most gifted engineers. But he and Hubbard already had been butting heads over planned impact tests that would fire pieces of foam at samples of the shuttle's tiles and RCC panels in an attempt to reproduce the damage seen on Columbia. Roe wanted a gradual ramp-up in tests before recreating the strike seen during launch. To complicate matters, Marshall engineers continued to insist the foam chunk was smaller than everyone else had estimated. Hubbard wanted to cut to the chase and immediately test a larger piece that reflected the consensus size. The CAIB had a deadline for completing its work. The spat got so heated that investigators inserted a text box explaining "the ongoing and continually unresolved debate" on page 78 of their final report.

"I suggested what we needed to do was a more methodical engineering approach," Roe explained later. "We had a range of sizes and a range of velocities that we thought it could be. We would . . . get the best data if we started small and slow and worked up to fast and big. But Hubbard would have nothing to do with that. Hubbard wanted his size and his velocity shot first, and that is what we had to do."

In addition to Ham's and Roe's ousters from the investigation, other changes were in the works. On April 19, the news broke that Dittemore would be stepping down as shuttle program manager later that summer to take a job in industry. He had made the decision before Columbia's launch, but had been waiting until the flight returned to announce his departure. He would be replaced by veteran NASA manager Bill Parsons. Additional changes were announced on July 2. Ham's position was largely eliminated and split up among several managers. She had been transferred to a job as an assistant to engineering director Frank Benz. Wayne Hale was promoted and called back to Houston to become Parsons' deputy and chair the Mission Management Team. Lambert Austin was replaced by engineering manager

John Muratore as head of Johnson's shuttle integration office. Veteran engineer Steve Poulos took over many of Roe's responsibilities, and Roe was reassigned to Langley to help set up a new, independent engineering safety office. Kennedy launch manager Ed Mango replaced Petete in the orbiter project office. The latter move was lumped into a NASA press release on the shakeup, although Petete's transfer had been planned well before Columbia's flight.

The announcement clearly was designed to give the impression of a housecleaning in the shuttle program at Johnson. The drumbeat for personal accountability slowly had been building on Capitol Hill and in the media. Privately, some at Johnson were outraged about the moves, especially the treatment of Ham, whom many considered a scapegoat.

In the weeks following the accident, Ham had been a lightening rod for criticism of the Mission Management Team's poor handling of the foam issue. She had been vilified in the press and portrayed by some investigators as a cold, uncaring bureaucrat who tolerated no dissent. The talented engineer whose career had skyrocketed up through the ranks suddenly found herself the embodiment of everything wrong with the agency's culture.

"If you ask, 'Do I take charge of a meeting for which I'm responsible?' the answer is yes, and I think it's necessary," Ham said later. "Do we make decisions? Well, anyone in program management makes them, and a lot of them are risk trades, and we make them every day. In the workplace, I'm not nurturing. I'm not sweet. I'm not mothering at all. But I don't think anyone expected that of me. I couldn't have been a flight director or in program management if I were that way."

The attacks mounted. One magazine article published 10 months after the accident went so far as to discuss her appearance and supposed habit of "wearing revealing clothes" without ever speaking to her. Ham was devastated.

"It's kind of funny," she said months after the accident, her tears reflecting a different feeling. "I probably went 10 years without hardly ever crying. But since February 1, it's like every day."

The obsession with Ham by the media, Congress and even some at NASA headquarters masked a bigger truth that Gehman and other investigators already understood. It was clear to the CAIB that Ham and mission managers had made bad decisions. They had failed to dig deeper, as good managers should, or think outside the box. They had, whether intentionally or inadvertently, stifled communications. In short, they had blown it. But the Mission Management Team wasn't responsible for the death of Columbia's crew.

"We felt very, very strongly that it isn't the failure to obtain on-orbit imagery that caused this accident," Gehman said later. "The failure that caused this accident is 15 years of starving and squeezing research and development, of living with violations of the Level 1 specifications, of having 3,000 waivers and not having a robustly funded engineering section that was methodically working through those waivers trying to eliminate them. That is what caused this accident. That, Linda Ham is not responsible for."

Another board member put it this way: "All of the decisions had been made all those years about foam not being a problem. That set up a game of musical chairs, and she just happened to be standing there when the music stopped. It could have been a previous flight. It could have been the next flight."

The building tension between Gehman and O'Keefe soon was working its way down through the ranks. The board was becoming increasingly frustrated with NASA's slowness in replying to requests for information. At first, there were isolated incidents of open rebellion as some shuttle engineers simply wouldn't respond. After the first couple of weeks, however, there were few cases of outright refusal. Instead, investigators occasionally would get more subtle resistance, or "pushback," in areas like the foam impact tests.

Board members occasionally wondered if anybody at NASA was listening. Investigators were stunned to see Ham take part in a joint NASA-CAIB briefing after she had been removed as head of the Mishap Response Team. Eyebrows arched even higher at an Aug. 6 CAIB executive session when Gehman and Barry briefed other board members on their official interview with O'Keefe. The administrator had dismissed some important issues—budget pressures and a freeze on shuttle upgrades—as "folklore" and "urban legend." Minutes from the executive session noted, "This is disturbing and indicates a different 'mental plot' from the board." O'Keefe had e-mailed the board an invitation to join him and other senior NASA officials for an informal farewell dinner the week of the report's release. The board sent back a note saying that following the report's release, they would agree to an interactive meeting with agency officials at NASA headquarters. They could have dinner afterward.

The feelings of frustration were mutual. CAIB investigators were asking tough questions during interviews with NASA officials. Many in the space agency felt their interviews were more like interrogations, with investigators having their minds made up before they entered the room. Others suggested that certain people with stories that contradicted the CAIB's working hypotheses, particularly in the area of how mission managers performed, never were called to testify.

"They [investigators] came in almost with an attitude," Johnson manager Paul Shack said. "They said, 'NASA's culture is broken, and we are here to fix it.' That's the way they started the interview. That was the first interview, and in my second interview, they kept asking the same questions from different directions, trying to get a different answer."

Even some board members later would confess that a few investigators occasionally got carried away.

"Most of the investigators had a pretty light touch," Wallace said, "but not all of them. We actually had to throttle a couple of them a couple of times."

Diggins wasn't one of those who had to be throttled. However, the veteran FAA investigator noted that it wasn't always easy to get the facts from reluctant witnesses.

"People will answer the question they feel comfortable answering, not necessarily the question that was asked," he said. "We were seeking information . . . Just because somebody says something doesn't mean it is necessarily so."

While shuttle managers at Johnson viewed the investigators with varying degrees of resentment, they were alternately bewildered and outraged that senior managers at NASA headquarters appeared to be held to a different standard. When the board's final report came out, they were livid, as one Johnson manager put it, that "the CAIB didn't lay a glove on them." Ham had been excoriated for not pursuing the photo request and for the manner in which she had presided over the

MMT meetings at Johnson. At NASA headquarters, Bill Readdy also had failed to pursue photos and had presided over the fateful FRR that accepted Marshall's flimsy rationale on why the tanks should continue flying after Atlantis' foam loss. Yet Readdy, who had offered his resignation to O'Keefe the day of the accident, was barely mentioned in the report.

Months later, recalling the FRR, Readdy offered this explanation: "We had every expectation that the information presented was accurate, certainly that everyone involved in the project, not only at Michoud but also at the Marshall Space Flight Center in the ET [external tank] project, thought they were presenting a very accurate picture. I don't remember hearing anyone discuss from the Lockheed Martin ET crowd or the folks there at Marshall that there were any concerns."

Even so, some of the board's staff acknowledged that responsibility for the accident went far beyond Houston.

"That happened in Challenger too and not only then, but in almost all cases where large organizations end up doing something bad," sociologist Vaughan said. "Typically, people that are higher up are not held accountable and the middle-level functionaries are the ones that are—the people who are most closely related to the decisions that were bad. . . . The ladder of responsibility needs to extend much higher than Linda Ham."

NASA managers outside the beltway pointed out the agency's top safety officials were located in Washington. Safety chief O'Connor had learned of the photo request and corresponded briefly with landing gear expert Carlisle Campbell on engineers' concerns. O'Connor also had been at Atlantis' fateful FRR, where he had argued with Jerry Smelser about whether the foam strike was a safety issue or an accepted risk. The former astronaut had seen images of the foam strike and had been briefed on the debris assessment team's findings. Like shuttle managers in Houston, he had done nothing.

No one, however, drew more sniping from outside Washington than O'Connor's predecessor as the safety chief at NASA headquarters, Fred Gregory. The 62-year-old former astronaut had served as the agency's top safety official for a nine-year period spanning three presidents, from 1992 to 2001. It was on his watch, shuttle managers insisted, that NASA's safety culture had all but evaporated. After O'Keefe took over as administrator, Gregory was promoted to NASA's number two spot, deputy administrator, in 2002.

A former associate administrator at NASA headquarters offered this blunt critique of Gregory.

"He definitely wasn't aggressive but was very passive," the official said. "I always left meetings with him wondering what he did. He never seemed to exhibit any leadership or initiative."

A senior shuttle manager at Johnson offered a briefer, more brutal assessment: "There was nobody home. Fred Gregory is a nice guy, but he didn't know shit."

Gregory, not surprisingly, disagreed, arguing his record contradicted the criticism.

"No one takes safety more seriously than I do," Gregory said. "I had probably 60 very successful flights during the nine years I was in the safety program, even though there were cutbacks and freezing of agency funding. What we did was adapt to the changing environment."

"I certainly can't comment on personal attacks that I guess have been levied at me," he added. "Sure, we have made some mistakes. I'm sure that when we look back we can probably say we should have done this instead of that. But our performance indicates the agency was doing a reasonable job during that period of time even though we were having major downsizing."

At an Aug. 5 press conference with Readdy and O'Connor, three weeks before the release of the CAIB's final report, Gregory downplayed what was by then widely publicized criticism of NASA's management culture. Two reporters grilled Gregory on the subject. He steadfastly refused to discuss management problems or even admit they existed.

Yet another reporter came at the question from a different angle, and finally Gregory replied that he believed most of the criticism of NASA's culture had originated with a single CAIB member. He added the board's findings would not be known until the report's release.

"It would be difficult for me to define to you what the 'NASA culture' is," Gregory said. "As I sit here, and I have three astronauts here, I suspect if you tried to determine what the culture of the three of us is, you would find there are three different cultures here. So that's why I have said, I have to wait and see what—if anything—is being written about culture before I can respond to your question."

Surprisingly, Gregory's name did not appear once in the CAIB's 248-page final report.

Dittemore recalled that shuttle managers at Johnson had no idea the agency's safety office was little more than an empty shell. They wrongly assumed the organization would provide the necessary checks and balances. It didn't.

"I did not realize how badly at the time our safety organization was functioning," Dittemore said. "The criticisms are fair. I think the [CAIB] report nailed it. I am so disappointed, beyond what I can tell you, that the safety organization that I relied on as my supreme check and balance against group think and nonadequate technical judgment and analysis was nonexistent. . . . In my mind, it was an absolutely critical failure."

O'Keefe decided not to go after senior managers at NASA Headquarters. Managers at Johnson remained irate months later that none of those changes had involved senior-level officials in Washington. The CAIB had decided early on to dodge the question of accountability. Determining what, if any, punishment was appropriate would be left to NASA. When the issue later came up in Congress, Gehman pointed out there was more than enough in the board's report to hold individuals accountable if that was what lawmakers and NASA wanted to do. But neither took action. The issue died. There would be no purge.

Vaughan argued that those involved already had suffered enough.

"I think, first of all, there isn't anything that you could do to these people that would punish them more than to know that they had this failure," she said later. "There is nothing that will ever remove the grief and sense of responsibility. Nothing. My interviews with the Challenger people [show] this just does not ever go away. In that sense, they are accountable. They become publicly accountable and thus lose their jobs and their prestige in the organization and their ability to command respect. So shifting them into other positions is an example of the recognition on the part of the organization that they can no longer lead in the same way that they did. That also is a kind of punishment, though it is not the kind of 'let the heads roll' that outsiders might call for. I personally think it is important that people stay on in their jobs and are not fired. . . . It keeps the institutional memory alive of what happened."

As the investigation continued, it became clear that intense schedule pressure to build the International Space Station was another factor in the accident. Senior officials at NASA Headquarters had created a sense of urgency where shuttle managers felt pressure to rationalize away problems and keep missions launching. Although Columbia's final flight was one of the few not headed to the station, it fell victim to the same pressures. There was a need to clear it from the manifest. Columbia's next mission was to the station. And because it was the only orbiter not equipped to visit the outpost, it needed to get back as soon as possible to have a docking system installed.

O'Keefe had come to NASA with the goal of bringing cost and schedule discipline to an organization that desperately needed it. The space station program essentially was on probation. It had to prove it could live within its means and stay on schedule before a decision would be made to expand its crew from three people to six or seven. That expansion was critical to meeting the outpost's research goals and accommodating NASA's international partners. In the minds of many program managers, however, the all-out push to meet a target date of Feb. 19, 2004, for completing the first phase of the station's assembly had gone too far. The date was an arbitrary deadline. As processing issues ate into the schedule, workers were assigned to come in on holidays and weekends. Testing was scaled back. To illustrate how hell-bent NASA headquarters was to stay on schedule, a computer screensaver was sent to program managers that showed the days, hours, minutes, and seconds ticking down to the so-called core complete deadline.

"The thing that concerned me," one worker told investigators, "is I wasn't convinced that people were being given enough time to work the problems correctly."

To keep the station on track, O'Keefe had brought in retired Air Force Gen. Mike Kostelnik to help crack the whip. Kostelnik had an impressive record that included commanding the Air Force's Development and Test Center and Air Armament Center at Eglin Air Force Base in Florida. The former test pilot had applied to be an astronaut three times between 1978 and 1982, but never was selected. Kostelnik had met O'Keefe in 1996 during a leadership development course at Syracuse University. He was named to the newly created post of deputy associate administrator for the station and shuttle programs at headquarters in May 2002, making him Readdy's deputy and Dittemore's boss.

It all was part of a move to shift management of the shuttle and station programs back to Washington from Houston. The Rogers Commission had urged that overall control be moved to Washington in 1986 after the Challenger accident exposed turf wars between the NASA field centers. Things had gone back to the pre-Challenger way of doing business in 1996 under the leadership of former NASA administrator Dan Goldin and powerful Johnson director George Abbey. The move prompted O'Connor—then the headquarters director of the shuttle program—to resign from NASA in protest. O'Connor had come back in June 2002, a month after Kostelnik's hiring, to replace Gregory as safety chief. Bringing Kostelnik and O'Connor aboard showed just how serious O'Keefe was about running the shuttle and station programs from Washington. Kostelnik's hire also sent the message that O'Keefe intended to stick to the schedule.

"Clearly, program management is all about cost, schedule and technical performance," Kostelnik said later. "You cannot manage or have a serious program that doesn't have those elements. So for somebody to presuppose that schedule should not be important, that would be a mistake. Now, cutting things, and in particular safety, to meet an arbitrary schedule—that also would be a mistake."

It took Kostelnik little time to alienate program managers at Johnson. During a management meeting in Galveston, Texas, three days before Columbia's launch, Kostelnik declared there was "a new sheriff in town" and the station schedule would not slip on his watch. In front of a room filled with senior managers, he pointed at Dittemore.

"He was very clear and very vocal about saying that he was the program manager for both shuttle and station," said a Johnson shuttle manager who attended the meeting. "He said, 'You just don't get it. You haven't gotten it yet that I am the program manager.' I went out and had a discussion with Ron [Dittemore] after that and said, 'Does this man realize the audience he is speaking to?' I felt he was demeaning and insulted the audience at a kickoff meeting that was supposed to be motivational."

Others in attendance walked away with similar feelings. Kostelnik referred to the intense schedule push as "leaning forward in the saddle," a term that elicited snickers and derision at Johnson, as did his desire to be referred to as "general." His not-so-subtle management style led to increasingly tense relations with Dittemore and other Johnson managers. Many later would speculate that the shift of shuttle program management to Kostelnik at headquarters was a major factor in Dittemore's resignation.

"[Kostelnik] expected us to respond to him like a bunch of privates," one Johnson manager said. "He would push back, and basically insulted them [senior managers] on a personal level. It was kind of hard for me to listen to. He is supposed to be leading us and motivating us, and I felt like he had the opposite effect."

From the headquarters perspective, each NASA field center had its own independent culture and considered itself the center of the universe. Johnson was the worst. Integrating all of the disparate parts into "one NASA" with control centralized in Washington was essential to O'Keefe's vision of bringing cost control and efficiency to the agency as a whole. Kostelnik was determined to do that.

"The system was, in reality, built by a very loose assembly of very strong centers," Kostelnik said. "So part of the Galveston meeting was trying to get the people in the program to think of themselves as the human spaceflight program or the space shuttle or ISS [International Space Station] program . . . not the Johnson piece of that, or the Marshall piece of that, or the Kennedy piece of that."

In the eyes of the CAIB, it all added up to a largely dysfunctional agency. NASA headquarters officials told investigators that concerns about station schedule pressures never made it to Washington. Evidence collected by the board suggested that simply wasn't true.

"NASA headquarters claimed nobody ever asked, and if they had asked, they would have delayed it [the deadline]," Gehman said. "We have dozens of people who walked in with the viewgraphs that they used showing that they did go to NASA headquarters and asked for relief from the [first phase] completion date and that schedule pressure was a problem, etc."

The budget and schedule pressures were nothing new. The more the CAIB dug, the clearer it became that many of the agency's problems were rooted in the past. In fact, most of the cultural issues the board identified had originated or worsened during the tenure of O'Keefe's predecessor as NASA administrator, Dan Goldin.

Goldin was one of the more colorful and influential figures in the space agency's history, serving the longest single stretch of any NASA administrator, from April 1992 to November 2001. Formerly the manager of defense contractor TRW's Space and Technology Group, the workaholic New Yorker was part rocket scientist, part tent-revival preacher—a self-styled "agent of change" who wore his trademark cowboy boots to black-tie dinners. He was O'Keefe's polar opposite. O'Keefe was the consummate manager, a level-headed pragmatist who cared more about reining in cost overruns than exploring the distant reaches of the solar system. O'Keefe never would be considered a visionary or a rocket scientist. Goldin passionately and genuinely believed in venturing out into the universe and understood the engineering details of NASA's most complex systems. But O'Keefe's even-tempered management style escaped him. By the time Goldin left NASA, the agency's financial records were so badly fouled up that a team of outside accountants took months to sort them out. The station program was chronically overbudget and behind schedule. His "faster, better, cheaper" approach to interplanetary probes led to spacecraft being built on the fly without the money needed to ensure success. The program resulted in embarassing setbacks, including back-to-back failures of two Mars missions in 1999 that generated enormous negative publicity.

Nor could the men's demeanors have been more different. Goldin's quick temper and tongue lashings were legendary among subordinates. In fact, his mercurial behavior earned him a nickname at NASA headquarters: Captain Crazy. His reputation followed him after he left the space agency. A deal for Goldin to become president of Boston University fell apart in October 2003 amid concerns about his management style.

"Dan stayed on too long," an associate administrator at NASA headquarters said. "He truly was a change agent. He pissed people off and rocked the boat, sometimes for good cause, sometimes for no reason. He probably should have left after four or five years. . . . He was not good at putting people in the right spots. He let Abbey and Gregory stay on far too long. . . . If you could somehow combine Dan Goldin and Sean O'Keefe, you would have the perfect NASA administrator."

The agency's "Goldin years" had been marked by unprecedented, sometimes tumultuous change. Cost-cutting was in vogue in the 1990s as the Clinton White House began an all-out effort to balance the federal budget. During that time, NASA experienced more than its fair share of cuts. The shuttle program was particularly hard-hit as resources were diverted to the new space station project. In 1993, NASA spent $4.02 billion on the shuttle program. By the time Goldin left in 2001, that figure had plunged to $3.12 billion (after dipping as low as $2.91 billion in 1998). The reduction represented a drop of about 40 percent in the shuttle program's purchasing power, compared to a 13 percent decline for the agency overall.

Goldin tackled the downsizing with a vengeance. He made no apologies for slashing areas he considered bloated and wasteful.

"When I ask for the budget to be cut, I'm told it's going to impact safety on the space shuttle," the board quoted Goldin as telling a 1994 NASA gathering. "I think that's a bunch of crap."

Others, like astronaut Ross, saw the cuts very differently. Smaller budgets meant job reductions, which, in turn, meant fewer people watching safety.

"It was Mr. Goldin who was saying, 'Thank you, sir. May we have another cut please?' " Ross recalled. "If I had been the administrator, a long time ago I would have said 'Over my dead body' and 'I am out of here if you do it.' I just don't understand why people, for eight years or whatever, continue to take that and say 'Boy, this is great.' "

The CAIB debated how extensively to chronicle Goldin's influence on the conditions that led to the accident. Logsdon, whose subgroup studied the issue, gave a briefing on the Goldin era to other board members at a May 21 internal meeting. According to minutes of the meeting, one chart noted that Goldin had so much stature in Washington he was praised by both Al Gore, the Democratic vice president, and Newt Gingrich, the Republican Speaker of the House. But

Logsdon expressed his belief that, regardless, "Goldin feared for his job and therefore did not fight very hard for NASA's fair share of the budget."

Gehman and Logsdon personally went to interview Goldin for the CAIB. Goldin refused to talk on the record and would not allow the interview to be recorded.

"We could have said, 'Okay, thanks very much. I will put this down as a refusal,' " Gehman recalled. "But instead we decided to stay and talk with him."

Gehman declined to give details of the discussion, but when pressed, he did say tersely, "I didn't learn anything new."

Ultimately, Goldin's influence was covered in six pages of the CAIB's report in a section titled "Turbulence in NASA hits the space shuttle program." A hard-hitting initial draft was watered down slightly for the final version.

The CAIB had uncovered massive organizational and institutional problems within NASA almost everywhere it had looked. The board's final report would conclude that "management practices overseeing the space shuttle program were as much a cause of the accident" as anything that had happened to Columbia's left wing. The comparisons with Challenger were undeniable.

"The Board is quite convinced," Gehman would say later at a news conference, "that most accident investigations do not go as far as we did, in that most accident investigations find the widget that broke, they find the person in the cause chain closest to the widget that broke, require that the widget be redesigned or replaced and the person fired or retrained, and then call it a day. And they do not go far enough to find out why did this happen. The failure of that is that you really haven't fixed the problem which caused the problem. You really are setting yourself up for a repeat."

TEN

RE-ENTRY REVISITED

The most complicated machine ever built got knocked out of the sky by a pound and a half of foam. I don't know how any of us could have seen that coming. The message that sends me is, we are walking the razor's edge. This is a dangerous business and it does not take much to knock you off.

—Flight director Paul Hill

On March 19, members of a salvage team were walking a search grid near Hemphill, Texas. They were stretched out in a long line, looking at the ground to either side as they trudged along. The area had been searched once before, but orders had come down to search it again.

The team was looking for what amounted to a needle in a haystack: Columbia's Modular Auxiliary Data System, or MADS, recorder, which stored launch and re-entry readings from hundreds of sensors. The recorder, as well as Columbia's flight computers, voice recorder, film, and videotape—anything that could provide stored or recorded information—was on a "hot list" of high-priority items. The list was updated daily, and the MADS recorder was at the top.

Only Columbia, NASA's original shuttle, was equipped with such instrumentation. It had been designed to collect engineering test data during the ship's first flights, and as such, it was no longer maintained or even required. Even so, some 570 of the system's 721 sensors were still operational when Columbia took off. If the recorder had survived Columbia's breakup, the data stored on its magnetic tape could be crucial to helping investigators figure out exactly what had gone wrong during the shuttle's final minutes.

But no one knew where the recorder was or whether it had fallen to Earth in one piece. It had been mounted in one of Columbia's crew module instrument bays. Search crews had found the mangled remains of equipment that had been mounted on either side of the MADS recorder, but the black box in question had not turned up.

Every piece of recovered shuttle wreckage was entered into a huge database that included the latitude and longitude where it had been found. Engineers plotted the locations where all the other components from avionics bay 3A had been found. The experts studied the map, looking for a pattern. They found one. In view of where the other debris had been located, the MADS recorder, or the remains of it, should be somewhere in a specific area near Hemphill.

It was an educated shot in the dark. No one was optimistic.

"The boxes on either side of it were 50 miles away and ripped to shreds," said Dave Whittle, the search director.

Search crews were briefed the night before. They were told what the recorder looked like and why it was so important. Art Baker, a Florida firefighter and a self-described man of faith, prayed for success. The next day, walking the line over hilly terrain, his prayers were answered when he saw a square metal box, painted black, about the size of two or three pizza boxes stacked on top of each other. It was lying flat on the ground, and appeared to be in pristine condition.

In one of those incredible breaks that defy explanation, the MADS recorder had somehow survived Columbia's breakup, fallen 37 miles and made it to the ground intact.

"It was just sitting there in a field; it was not caked in mud; it was not buried under ground; it was not busted apart," marveled NASA test director Steve Altemus. "Its tape reels looked like brand new."

Within hours, the MADS recorder was on Whittle's desk and, shortly after, on a jet to Imation Corp. in Minnesota, where data storage specialists carefully removed and cleaned the 1-inch-wide tape. Before any attempt was made to actually play back the recorded data, it was flown to Kennedy and copied. Only then was it shipped to Johnson, where a team of engineers was standing by to begin an exhaustive analysis.

"The recorder really buoyed everybody's spirits," Altemus said. "That was just huge. We all felt like, yeah, we're going to get this. We are going to figure this thing out."

By this point, they already knew a piece of foam insulation from the shuttle's external tank had broken free and hit the underside of the left wing. They already knew from analysis of wreckage and telemetry radioed back to Earth during the shuttle's descent that a breach of some sort had allowed super-heated air to eat its way into the interior of the wing with catastrophic results. They already knew from studying debris and video shot by amateurs along Columbia's flight path that the shuttle had been in severe distress, literally falling apart, all the way across the southwestern United States. They were still considering the possibility of a breach near the left main landing gear door. But they were beginning to suspect the impact had occurred closer to the front of the wing, possibly somewhere on or very near the leading edge itself.

Six weeks into the investigation, exactly what had caused the disaster remained a mystery. Many at NASA still did not believe the foam strike triggered the mishap and the MADS data held no direct proof either way. Resolving that question would require extensive tests to determine the impact velocity and the strength of the shuttle's tiles and leading edge panels. But the 28-track tape gave investigators the data they needed to piece together exactly where the breach must have occurred and how the super-heated air had worked its way into the left wing's interior.

"We had found enough debris that we were suspicious of the leading edge, but frankly we weren't sure that it wasn't some broken tiles around the left main landing gear door. That was still in play," said Frank Buzzard, director of the NASA task force supporting the Columbia Accident Investigation Board. "If it had been a shuttle other than Columbia, we would still be having this debate. The Columbia had that [MADS] instrumentation, and the Lord provided that box for us, and within three or four days of analyzing that data, everybody started to converge on what had to be [a breach in] the leading edge. So, we got over this denial stuff."

The MADS data served as a reality check. It had to match up with the damage seen in the recovered wreckage. If the recorded sensor data indicated a breach at one location, for example, physical evidence was expected to be in agreement. Data radioed to Mission Control during Columbia's descent also had to match up. If the size or the location of the breach could not explain the aerodynamic forces detected by the autopilot system, investigators would know they had problems with the failure scenario.

The recovered debris alone made a compelling case. More than 25,000 men and women from 270 local, state, and federal organizations ultimately scoured 2.3 million acres—700,000 acres on foot—looking for wreckage in a "debris footprint" that eventually measured 645 miles long and 10 miles wide. In Texas alone, the footprint exceeded 2,000 square miles. Search teams recovered some 84,900 individual pieces of wreckage making up about 38 percent of the space shuttle. The operation cost the Federal Emergency Management Agency, the coordinating organization, more than $305 million.

Heavier pieces of wreckage from the forward fuselage and aft engine compartment flew the farthest, crashing to Earth in eastern Texas and western Louisiana. One 800-pound main engine component hit the ground at twice the speed of sound and buried itself 12 feet deep. Another 600-pound engine fragment hit a golf course, blasting through the water table and creating a small pond. A golfer later complimented the course pro, NASA Administrator Sean O'Keefe recalled. The golfer said, "Gee, I really like that new pond you put over there on No. 6." The pro said, "What are you talking about, new pond? We don't have a pond on 6." And the golfer said, "Well, you sure as hell do."

NASA launch director Mike Leinbach, who spent two weeks in the field before returning to Kennedy to oversee the reconstruction effort, was called to the scene of one engine component, found buried in a dense forest. The force of the impact had splashed mud 20 to 30 feet high in the surrounding trees. As he walked away, a Louisiana Forest Service supervisor in front of him suddenly started crying. The two hugged, both now in tears. "I just wish there was more we could do for you," she sobbed.

Debris from the left wing was found farther west than debris from the right, indicating the left wing broke up first. But the left wing was heavily damaged long before its ultimate failure, and salvage crews found much less of it. The most westerly piece of recovered wreckage, found near Littlefield, Texas, was a single heavily charred tile that had been mounted just behind reinforced carbon-carbon panel 9 on the left wing. This was a critical find because it supported the idea the failure began in that region of the wing. Other material from that area had undoubtedly fallen to Earth farther west, but as 2003 drew to a close, the Littlefield tile remained the westernmost piece of confirmed shuttle debris.

Twenty-seven truckloads of wreckage were hauled to Kennedy between Feb. 5 and May 6. Each piece or component was cleaned, decontaminated, bar-coded, photographed and entered into a computer database. Seeing Columbia come home in boxes was traumatic for the men and women who had devoted their careers to servicing NASA's first shuttle.

The first load of wreckage to arrive at Kennedy included spherical oxygen and hydrogen tanks that had been part of Columbia's fuel cell system. Altemus recalled seeing a fuel cell engineer standing there just looking at this truck, all by himself. "I said, 'Hey, Tom, how are you doing?' And he just shook his head and walked away and I never saw him again."

Wreckage from Columbia's wings slowly trickled in. But the few charred remnants bore little resemblance to the gracefully curving wings that once carried Columbia and its crews home from orbit. Even so, that wreckage—or the lack of it—held one set of keys to understanding the disaster. Sir Arthur Conan Doyle's Sherlock Holmes once solved a case after realizing a dog that should have barked, didn't. By understanding what was missing from the wing, shuttle investigators hoped to trace the failure to its source.

It was a detective story worthy of Doyle's hero.

Shuttle wings are made of aluminum, the upper and lower surfaces separated by spars and trusses that form a boxlike internal framework. The main landing gear wheel well boxes are located toward the front of each wing, nestled up against the side of the orbiter's fuselage just behind the leading edge. Near the wheel wells, the wings are spacious enough for a person to easily crawl around inside, but the space diminishes toward the rear of the wing as the two surfaces come together. Openings in the spars running from the fuselage toward the wingtip allow wires to pass through and permit pressure to equalize in the wing's interior as the shuttle descends into the atmosphere.

Columbia's left main landing gear strut, a few fragments of the landing gear door and both heavily damaged tires were recovered in Texas. But none of the internal framework of the left wing, or any of its aluminum skin, was ever found. Aluminum melts at about 1,200 degrees. Shuttle design requirements call for insulating the shuttle's skin to prevent it from getting any hotter than 350 degrees during entry. On the top of the wing, where re-entry temperatures are lower, insulating blankets are used, along with heat-shield tiles located around the periphery. On the bottom of the wing, where temperatures exceed 2,000 degrees during the region of maximum heating, custom-fitted tiles are used throughout, each one coated with a black surface layer to improve its ability to dissipate heat.

None of the insulating blankets from Columbia's left wing was ever found. Heavily damaged tiles were recovered, but not nearly as many as were found from the right wing. Tests and analyses showed at least some of the recovered tiles apparently fell off because the adhesive used to attach them to the skin of the wing had broken down. Extreme heat obviously had made its way into the interior of the wing, warming it up from the inside out.

The goal of the debris reconstruction effort was to discover where the heat had come from. As more and more wreckage arrived at Kennedy, it became increasingly clear the breach must have been in the leading edge. But where?

Behind its protective insulation, the front of a shuttle wing is flat, made up of a panel of aluminum honeycomb material known as the leading edge spar. To give the wing its aerodynamic shape, and to protect it from the most extreme temperatures of re-entry, 22 RCC panels are bolted side by side on that flat front surface, creating a smoothly curving leading edge. So-called spanner beams, made out of a heat-resistant alloy called Inconel, provide rigidity. To seal the gaps between RCC panels, thin carbon-composite strips called T-seals are bolted in place to provide a smooth surface along the entire leading edge. Air is allowed to enter and exit the leading edge cavity through slender vents between the RCC panels and the upper surface of the wing to equalize pressure as the shuttle descends into the atmosphere.

The U-shaped interior of the RCC leading edge just in front of the forward spar is hollow. During re-entry, the shuttle's nose is pitched up 40 degrees, which subjects the lower halves of the RCC panels to the most extreme heating. The inner surfaces of the panels radiate some of that heat inside the cavity just in front of the main spar. The fittings used to attach the RCC panels to the main spar are protected by heat-resistant insulation that melts at 3,200 degrees.

To mount or remove RCC panels, which are numbered from 1 to 22 starting where the wing joins the fuselage, workers must have access to the attachment fittings. For this reason, a gap exists between the tiles on the upper and lower surfaces of the wing and the rear edges of the RCC panels. That gap is filled with rectangular, tile-covered "carrier panels" that are bolted in place as a final step in the assembly process. The result is a seamless transition from the RCC panels to the tiles protecting the rest of the wing. The only gap is the tenth-of-an-inch-wide slit providing the venting needed to equalize air pressure in the RCC cavity as the shuttle descends.

Whatever happened to Columbia had utterly destroyed this complex system.

Wreckage from Columbia's wings, fuselage, and nose section was laid out on a grid in the Reusable Launch Vehicle Hangar near Kennedy's shuttle runway. Left wing RCC fragments, attachment fittings, leading edge access panels, landing gear components, and other critical items were subjected to exhaustive analysis. The most critical RCC panels and attachment fittings—those numbered 1 through 13 and nearest the fuselage—were mounted on a full-scale clear plastic mockup of the rounded leading edge that allowed investigators to see each piece in relationship with its neighbors. It also allowed them to map out exactly where the heat went after it entered the leading edge.

Analysis of the wreckage helped investigators rule out several theories about what might have gone wrong and led them toward a specific RCC panel as the location of the foam strike and the initial breach in the wing.

"Our conclusion was that it was somewhere on the bottom acreage area of No. 8 based on just the wreckage alone," Leinbach said, "So we put on a presentation for the board. . . . We showed them our conclusion, and they were impressed. They said, 'OK, we accept what you say, but let's go independently come to that same conclusion.' "

At Johnson, engineers were working around the clock to do just that—develop a credible failure scenario based on transmitted and recorded data alone. A group led by entry flight director LeRoy Cain started out with 10 possible failure scenarios. With the MADS tape in hand, they were able to zero in on the same working scenario the debris analysis team was reaching.

"It was a huge challenge," Cain said. "We started out, really and truly, with about 10 different possible lists and we narrowed that down. . . . Not really until we got the MADS recorder data did we really say, 'OK, we are not talking about a breach of the main landing gear door. We are not talking about the gear coming down.' It was the MADS recorder that sealed the deal in that whole leading edge thing."

Using engineering drawings and photographs made during Columbia's most recent overhaul, they could precisely locate every sensor. They also knew exactly where the wires leading to each sensor ran as they exited the fuselage and fanned out through the wing. Many of those wires fed data to the MADS recorder. Others carried readings to Columbia's flight computers and, ultimately, to Mission Control.

Cain's team had the actual readings from each sensor and the exact time when data from each one were lost. Tests were run to determine how much time a plume of super-heated air would need to burn through sensor wiring and spar panels. Putting everything together, they, too, came to the conclusion the initial breach must have been on the lower half of RCC panel 8. What's more, they were even able to make an estimate of its size: 6 to 10 inches across.

But the MADS data gave investigators much more than the breach's presumed size and location. It also helped them map out how hot air burned its way inside the wing during entry. The in-flight destruction of the left wing had a noticeable effect on Columbia's aerodynamics, forcing the shuttle's flight computers to take corrective actions. Engineers analyzed those computer commands to determine the possible aerodynamic effects of such a breach. As it turned out, a hole in RCC panel 8 fit the bill. The MADS data also provided insight into what happened during the shuttle's final seconds, after contact from Mission Control was lost.

Another team under flight director Paul Hill carried out an extensive review of film and videotape shot by shuttle watchers along Columbia's flight path to determine precisely when major pieces of the spacecraft fell off. Hill's team also scoured FAA and military radar recordings to trace the larger pieces of wreckage to the ground in hopes of pinpointing them for search crews west of Texas. It was complicated, time-consuming work. But despite several good radar tracks, nothing was found farther west than Littlefield.

In the end, it didn't matter. Taken together, the independent lines of analysis made such a strong case the CAIB was able to conclude, without qualification, that the foam impact was the root cause of the accident; that the impact had knocked a 6- to 10-inch hole in the lower half of RCC panel 8 on the shuttle's left wing; and that a plume of super-heated plasma entering through that breach had destroyed the wing and triggered the destruction of the orbiter.

But it took three months to iron out all the details.

First, investigators needed a better sense of exactly what had happened during Columbia's launch. An image analysis group essentially repeated the work of the original Boeing-led debris assessment team to determine precisely the size and shape of the foam that had broken away from the tank, as well as its weight, angle of impact and velocity. The available camera views were enhanced by the National Imagery and Mapping Agency to bring out subtle details. Sophisticated computational fluid dynamics calculations were carried out to characterize the airstream around the shuttle as it climbed toward space and to analyze the possible trajectories of the foam.

The team concluded the foam broke away from the left bipod ramp 81.7 seconds after liftoff and hit the underside of Columbia's left wing two-tenths of a second later. The foam measured 21 to 27 inches long by 12 to 18 inches wide. It was tumbling at 18 revolutions per second. Before the foam separated, the shuttle—and the foam—had a velocity of 1,568 mph, about twice the speed of sound. Because of its low density, the foam rapidly decelerated once in the airstream, slowing by 550 mph in that two-tenths of a second. The foam didn't fall on to the leading edge of the left wing as much as the shuttle ran into it from below. The relative speed of the collision was more than 500 mph, delivering more than a ton of force.

Many shuttle managers later expressed surprise the foam slowed so rapidly. They intuitively believed the foam had decelerated less and as a result, the impact velocity had been lower. They were wrong.

Analysis of enhanced launch pictures showed the foam struck somewhere in a 20-square-foot area covering RCC panels 5 though 9. The bulk of the material disintegrated on impact, and a spray of particles was seen emerging from under the wing. Heavily enhanced views of the strike appeared to show some of the particles flowing back over the top of the wing, indicating the strike may have occurred more toward the front of the leading edge. But the data were inconclusive.

The MADS recorder was running during launch, and its readings were added to the working scenario. Two pressure sensors on the lower surface of the left wing, directly behind the presumed impact point, showed signs of possible contact with debris particles. A MADS temperature sensor mounted just behind RCC panel 9 showed an unusual increase five to six minutes later. While the increase did not prove a nearby breach was present, a detailed analysis showed it was consistent with a 10-inch hole in RCC panel 8.

But why did the foam come off in the first place? Shuttle engineers had long believed trapped air or nitrogen in the foam could liquefy before launch, turning into a gas as the shuttle climbed away and the tank warmed up. The sudden increase in pressure, the thinking went, could blow off pieces of foam. But CAIB member and physicist Doug Osheroff carried out experiments showing such "cryopumping," on its own, could not explain foam shedding. Other factors had to be involved.

During Columbia's launch, the shuttle encountered high winds at the same time the ship endured the flight's maximum aerodynamic pressure. This occurred 57 seconds after liftoff at an altitude of 32,000 feet, around the time the spacecraft was rocketing through the sound barrier. The wind was hitting the shuttle from the left and pushing it to the right slightly, causing subtle changes in the flow of air around the left bipod ramp. But tests and computer modeling showed wind shear, by itself, could not account for the loss of the foam.

Other factors also were considered, including the vibration levels experienced by the shuttle, the effects of liquid oxygen sloshing about in the top of the external tank, the results of shuttle steering in response to wind shear during the region of maximum aerodynamic pressure, and the effects of a slight thrust mismatch between Columbia's two solid-fuel boosters. Again, no single factor emerged that could explain the foam loss.

"The precise reasons why the left bipod foam ramp was lost from the external tank during STS-107 may never be known," the CAIB concluded. "The specific initiating event may likewise remain a mystery."

The loss probably was the result of a combination of factors, including inconsistencies in foam application and undetected manufacturing defects as well as aerodynamic stresses, vibrations and extreme sound levels and temperatures.

Whatever the cause, the foam clearly came off and it clearly hit the shuttle. But what damage did it do? During the mission, the debris assessment team concluded foam wouldn't have hurt the RCC panels or caused major tile damage. But the RCC panels were never designed to withstand debris impacts, and the CAIB quickly discovered NASA had no test data showing whether the panels could stand up to the sort of impact that had occurred during Columbia's launch.

Even so, many at NASA, agreeing with program manager Ron Dittemore's earlier assessment, still did not believe the launch-day foam strike was at the root of the disaster. If it was, many thought some other factor must have been involved. The CAIB decided to settle the matter once and for all.

NASA and the board already had agreed to build a full-scale mockup of the area of the wing leading edge where investigators suspected the foam might have struck. The rig was shipped to the Southwest Research Institute in San Antonio, Texas, where a powerful nitrogen-gas-powered cannon was available. The gun typically was used to fire rubber "chicken simulators" at various aircraft components, like cockpit windshields, to test their ability to withstand bird strikes and similar collisions. In this case, the gun would blast chunks of foam at lower wing tiles and one or more RCC panels on the leading edge mockup.

The first test runs were designed to find out what the foam would have done to the tiles on the underside of Columbia's left wing. To duplicate the conditions that occurred during launch, a briefcase-sized foam block with a volume of 1,200 cubic inches was fired at a panel of tiles, striking with a velocity of more than 500 mph. As tile expert Calvin Schomburg had predicted during the flight, the impact caused no major damage. The RCC panels were another matter.

This time, the leading edge mockup was used. After initial test shots, foam was fired at a RCC panel taken from the number 6 position on Discovery. The panel was dislodged slightly and suffered several cracks. The results were dramatic and showed a foam strike could, in fact, damage the leading edge. But investigators needed to find out what such an impact would do to RCC panel 8, the one believed to have been damaged during Columbia's launch. Panel 8 was larger than panel 6 and was shaped differently.

It was at this point that board member Scott Hubbard and Ralph Roe argued about how to conduct the tests. In the end, the board took over the impact testing and left NASA with no choice about how to proceed.

On July 7, a 1,200-cubic-inch block of foam weighing 1.67 pounds was fired from the cannon at 530 mph. Because the real foam chunk was rotating as it left Columbia's external tank, the gun's rectangular barrel was "clocked," or rolled, 30 degrees to one side to slightly increase the impact energy. The target was RCC panel 8 taken from the shuttle Atlantis.

To a crowd of reporters and NASA officials looking on from a safe distance away, the foam's passage was too swift to see. But the results were instantly obvious. A ragged 16-inch-wide hole suddenly appeared on the lower side of the panel. The crowd gasped at the violence of the impact. Hubbard was ecstatic.

"We believe we have found the smoking gun. We believe we've established that the foam block that fell off the external tank was, in fact, the most probable cause, the direct cause of the Columbia accident," Hubble told reporters. "I've now got a direct connection between foam shedding creating a hole that's the same order of magnitude as what must have been there when Columbia came home on Feb. 1."

The impact blasted a 225-square-inch hole in the panel, big enough for investigators to stick their heads inside. Several large panel fragments were blown into the cavity directly behind the leading edge.

One of the enduring mysteries of the investigation had been military radar data showing an unidentified piece of debris had separated from Columbia the day after launch. Hubbard now said the "flight day 2 object" probably was a large section of the blasted RCC panel that worked its way free in the weightlessness of orbit.

That was mostly speculation. But with the impact test, no one harbored any lingering doubts about whether "the foam did it."

"I was surprised. I was very surprised," Hubbard told reporters. "As a physicist conducting a test, I feel gratified that after months of work we were able to demonstrate this connection between the foam and the damage. But I know it was a source of tragedy, so that makes me feel very sad. This whole six months, we've constantly been reminded by pictures of the seven lost astronauts what this all means."

With the dramatic foam shot at RCC panel 8, all the pieces of the puzzle were finally in place. There was little doubt about what had doomed Columbia and its crew. A second-by-second timeline of the final working scenario provided a gripping account of the shuttle's final minutes.

At 8:44:09 a.m. Eastern time on Feb. 1, 2003, Columbia was a half-hour from home. The shuttle had just dropped below an altitude of 76 miles, slipping into the discernible atmosphere 900 miles northwest of Honolulu. The shuttle's nose was pitched up, its wings were level, and its velocity was 24.6 times the speed of sound, still roughly 5 miles per second. Astronaut Laurel Clark, aiming a videocamera through an overhead window, commented that she was starting to see "some good stuff" outside as hot gas enveloped the shuttle's verticle tail.

During re-entry, the shuttle compresses the thin air in front of it, creating two shock waves. Those shock waves intersect around RCC panel 9, subjecting panels in that area to the most extreme heating. But the compression of the air in front of the shuttle forms a so-called boundary layer, a region just a few inches thick that resists further compression and acts as a natural insulator. Just a few inches away from the leading edge, just beyond the boundary layer, atoms are torn apart and temperatures can exceed 10,000 degrees. But the boundary layer keeps temperatures on the leading edge RCC panels at around 3,000 degrees.

A smooth surface is essential for the boundary layer to form, and is crucial to a shuttle's survival during the plunge to Earth. If the boundary layer is disturbed for any reason, its insulating effect can be compromised by high-temperature turbulence, subjecting the shuttle's tiles and RCC panels to much more heat than they were designed to handle.

"You have this big, massive thing coming through, slamming into the molecules, more and more as you get farther into the atmosphere, and that collision basically disassociates these molecules into their constituents," CAIB member Jim Hallock explained. "And that's where all this heating is coming from. [The shuttle] uses aerodynamic braking, basically. So you've got to get down to where the molecules are, to slow it down. That's really what's slowing it down, running into these molecules and dissipating a lot of the energy of speed into heat. . . . Once this boundary layer forms, it really protects you pretty well."

As the Columbia astronauts chatted about the light show outside, ever more air molecules were shooting into the hole in the lower side of RCC panel 8 and slamming into the insulation protecting the panel attachment fittings, swirling through the cavity and spreading out to either side. At that altitude, the effect was small. But the shuttle was descending, and the air was getting thicker with each passing second. The initially subzero temperature in the RCC cavity was beginning to climb.

Less than two minutes after entry interface, the view out the forward cockpit windows had changed from the black of space to a diffuse orange glow as air molecules and atoms were ripped apart just outside the boundary layer protecting the orbiter's underside, its nose cap, and the leading edge panels.

The recovered crew cabin videotape ended less than three and a half minutes after entry interface and a full minute before any of Columbia's sensors detected a problem. The astronauts were clearly unaware of what was going on just a few feet behind them. They had 15 minutes to live.

With Columbia in a 40-degree nose-up orientation, the plume entering the breach in RCC panel 8 was aimed at the upper attachment fittings and insulation. The insulation began melting. Molten globules blown back onto the inner surface of the RCC panel would provide tell-tale clues seen later in the reconstruction hangar at Kennedy.

As the insulation burned, the front face of the left wing's aluminum honeycomb forward spar—the only barrier between the plume and the interior of the wing—began heating up.

At 8:48:39 a.m., just four minutes and 30 seconds after Columbia dipped into the atmosphere 1,700 miles to the west, a sensor mounted behind the forward spar, near the point where RCC panel 9 was bolted to the other side, measured an unusual increase in stress. It was the first indication of a problem, and it confirmed beyond any doubt that Columbia re-entered the atmosphere with a pre-existing breach in the leading edge.

The MADS sensor, called a strain gauge, showed the aluminum of the spar was expanding as the relentless heating continued, causing stresses to build up in the honeycomb panel. The spar began softening. From the strength of the readings, investigators later calculated the sensor must have been within 15 inches of the point where the plume, or a portion of it, was hitting the forward face of the spar. That fit the profile for a plume entering through a breach in the lower half of RCC panel 8.

The shuttle was 54 miles up and dropping fast. Twenty seconds after the initial strain gauge readings, a MADS temperature sensor mounted on the front side of the spar between RCC panels 9 and 10 began measuring an unusual temperature increase. This was the only MADS sensor in the leading edge cavity and it was heavily insulated.

Investigators later determined the hole in RCC panel 8 must have been 6 to 10 inches across to produce the kind of hot gas flow in the RCC cavity needed to generate the observed readings from an insulated sensor many inches away.

About a minute after the spar strain gauge first detected unusual stress—five and a half minutes after entry interface—the shuttle's flight computers ordered a turn to the right. Up until this point, the shuttle had simply been falling into the atmosphere along its orbital track, wings level, nose up and pointed straight ahead. Now, the ship's flight computers began actively guiding the shuttle toward Kennedy's runway. The shuttle's nose smoothly swung 80 degrees to the right.

Less than 20 seconds after the maneuver, sensors mounted on Columbia's left rear rocket pod measured an unusual change in temperature. Instead of steadily climbing as one would expect during a normal entry, the temperature rose more slowly. Wind tunnel testing would later show some of the hot air blasting into the RCC cavity was exiting through the vents on the upper surface of the wing, carrying thin clouds of metallic vapor with it from melted insulation. The unusual plume interfered with the normal flow of air around the spacecraft, slowing the expected temperature increase on the left fuselage.

Later, as the breach in the leading edge worsened, the flow above the wing would change direction again and cause temperatures on the fuselage to rise sharply as a spray of melted Inconel was deposited on the ship's outer skin. Metallic vapor around the shuttle also would interfere with communications throughout the descent, in much the same way metal chaff dropped from fighter jets confuses enemy radar systems.

The firestorm inside the RCC cavity was rapidly increasing in intensity. The boundary layer around the leading edge breach was severely disrupted, and the flow of super-heated air over the lower surface of the wing exposed the protective tiles there to much higher temperatures than they were designed to withstand. Insulation and RCC panel support fittings behind the breach continued to burn away.

Within a few seconds of 8:52:16 a.m.—the exact time is unknown—the deadly plume entering through the breach in RCC panel 8 burned its way through the forward wing spar and into the interior of the wing. As it did so, it cut through sensor wiring attached to the interior side of the spar and began heating the internal spars and trusses forming the wing's aluminum framework. The outboard side of the main landing gear wheel well box was just 4 and a half feet away. Over the next 10 seconds, data from 17 of 18 leading edge sensors were lost as the spar breach widened from a small hole to a gaping wound. Ground tests later would show a plume similar to the one believed to be raging in Columbia's leading edge cavity could enlarge a 1-inch hole in an aluminum spar panel to 5 inches across in just 13 seconds.

The shuttle was still 300 miles from the coast of California. The crew still had no idea anything was wrong. The sensor wires that were destroyed when the forward spar was breached were sending signals to the MADS recorder. The data were not included in the steady stream of information being transmitted to Mission Control. Based on the timing of sensor failures and the known locations of the wires on the backside of the spar, the hole was at least 9 inches across in its initial stages and probably grew in size as the entry continued.

"I think you can make a case that you basically lost the spar from behind [RCC panel 8] all the way over to behind 9," said Air Force Lt. Col. Pat Goodman, a CAIB investigator. "And there's no reason really to think it doesn't keep going. When you match up with the debris evidence, there are no spanner beams from that area whatsoever; there is no spar from that area whatsoever; there are no carrier panels from [behind RCC panels] 9, 10 and 11. There's no RCC from 9 hardly at all.

"The only thing you can say is it was a minimum of nine inches," he said. "When you consider that we found really nothing of RCC panel 9, it just leads you to believe you eventually end up with some rather large hole [in the forward spar]."

As far as Cain and the rest of the flight control team knew, Columbia was descending normally. Husband had not spoken to Mission Control in more than 15 minutes. He said nothing now. As far as he or anyone else knew, Columbia would be back on the ground at Kennedy in just 24 minutes.

But with the boundary layer disrupted, the temperature of the atoms and molecules blasting into the wing probably exceeded 8,000 degrees near the leading edge breach itself. Within 14 seconds of the spar burn through, the plume began destroying three bundles of electrical wires running along the outboard wall of the main landing gear wheel well. Those wires carried data from sensors located throughout the left wing, including the hydraulic system used to move the elevons as required to keep the shuttle on course. Hot gas began flowing into the wheel well through vents around landing gear door hinges. The outboard wall of the wheel well now began heating up.

At 8:52:17 a.m., the first unusual sensor reading flashed on a computer screen in Mission Control: a slight increase in temperature in the hydraulic fluid running through a brake line leading to the left main landing gear. At the same time, a strain gauge mounted on the front of the wheel well box began showing signs of elevated temperatures, data that was routed to the MADS recorder.

Columbia's left wing was burning up from the inside out. The ragged edge of the breach in RCC panel 8 was beginning to affect the ship's handling as it dropped ever deeper into denser air. Twelve seconds after the brake line temperature reading showed up in Mission Control, the shuttle's flight computers noticed the effects of the damage for the first time as a force, or drag, began pulling the shuttle's nose to the left. After assessing the data for a few seconds, the computers sent commands to the elevons on both wings to push the shuttle's nose slightly to the right to balance it out.

Wind tunnel testing would later show a breach in the lower part of RCC panel 8 would cause enough aerodynamic disturbance to produce the same levels of drag detected by Columbia's computers. But on the flight deck that day, Husband and McCool remained oblivious to their ship's ongoing destruction. They might have noticed the elevon movement on their forward computer displays, but the adjustments were small and would not have caused concern.

Columbia finally crossed the coast of California north of San Francisco at 8:53:28 a.m. at an altitude of 45 miles and a velocity of 15,800 mph. Only 9 minutes and 19 seconds had passed since Columbia began its descent through the atmosphere. But by the time the shuttle crossed the coast, Columbia was in severe distress.

The spars and trusses inside the left wing immediately behind the forward spar were softening in the heat. The aluminum skin of the wing was warming up, and the adhesive used to hold tiles and insulation blankets to the outer surfaces was beginning to break down. Additional sensors failed as wiring melted and temperatures began rising sharply along the left fuselage. The flow of air over the wing changed yet again because of the worsening leading edge breach.

Within 10 seconds of Columbia's coastal crossing, virtually all of the sensors in the outboard regions of the left wing had been lost. Some of the spars and trusses giving the wing's upper and lower surfaces their rigidity likely had been lost as well, which allowed the skin to flex slightly.

"Basically, you're creating kind of an oven in there," Goodman said. "You keep pumping in this hot gas, and as you fill that up, now you start to drive up your internal temperature. So you might have direct [extremely hot] plume impingement on some of those trusses. But once you start getting that temperature above, say, 350, 400, 500 degrees, those aluminum trusses may be physically in place, but they're not carrying any load any more because they're softening."

Scores of amateur shuttle watchers in California and Nevada had gotten up before dawn to watch Columbia's fiery descent. Even first-time observers were struck by the appearance of the shuttle's plasma trail. The super-heated air left in the shuttle's wake that glowed in the dark sky like a phosphorescent contrail. Beginning just 20 seconds after the shuttle crossed the coastline, many observers saw what appeared to be debris falling away from the orbiter, sparkling in the plasma trail like burning embers. They almost certainly were seeing tiles and insulation blankets released from the left wing, either because the adhesive had failed or because the skin of the wing was buckling slightly—or both.

Photographer Gene Blevins, watching from the Owens Valley Radio Observatory near Death Valley, recalled his surprise.

"I was looking straight north and I could see little red, like lava chunks, falling off from it, flaking off from it and then getting sucked back into the plasma trail," he recalled.

The plume shooting into the wing from the front spar breach may have burned a hole through the upper skin of the wing during this period, perhaps at the same time that many observers on the ground saw a bright flash.

By 8:54 a.m., just 32 seconds after Columbia had crossed the coast—and just a minute and 44 seconds after the forward spar had been breached—the outboard wall of the left main landing gear wheel well began melting. A scant 11 seconds after that, the shuttle's flight computers detected another change in the way Columbia's flight path was being affected by the increasing damage to the left wing. Before this moment, two forces were at work. One was acting to pull the shuttle's nose to the left—drag—and the other was acting to roll the spacecraft to the left as well, the result of decreased lift from the crippled left wing.

Now, quite suddenly, the force trying to roll the shuttle over reversed itself. It was as if the left wing had suddenly gained additional lift. The ship's flight computers instantly responded, adjusting Columbia's elevons yet again to exactly counteract the two unwanted motions.

The shuttle stayed on course. Husband and McCool may have noticed the elevon movements as the autopilot responded, but again, they made no attempt to contact Mission Control for an explanation. In all likelihood, they still believed the entry was proceeding normally.

The increased lift initially puzzled investigators until they pieced together the plume's path through the wing's interior. The melting of the support spars and trusses just behind the forward spar caused the upper and lower wing surfaces to lose their rigidity. The lower wing, which was directly affected by the increasing pressure of the air, bowed inward, forming a depression. It started out small, but as the seconds ticked by and the wing's interior got even hotter, it grew alarmingly. Over the next five minutes, the depression probably grew to some 20 feet in length and 4 feet in width, a concave area more than 5 inches deep. Wind-tunnel testing and computer simulations later showed such a depression could explain the reaction of Columbia's flight computers.

In any case, within two seconds of the change in the shuttle's tendency to roll to the right, observers on the ground saw a large piece of debris fall away. Analysis of photographs and videotape shot by observers along Columbia's flight path show the debris probably weighed some 190 pounds. It may have been a large section of the lower wing, its loss possibly a direct cause of the change in the wing's lift.

"This far into the scenario it's a lot of supposition," Hallock said. "At the beginning, we've got some pretty solid facts. . . . From now on, we're really surmising what we think has happened. I'm one of those who believes, and most of us do, that the trusses and everything are softening so much, the upper and lower surfaces are trying to come together."

Goodman agreed, saying, "If you think about it, there's nothing holding the lower wing apart from the upper wing with the exception of the RCC panels. We've burned through the spar; that gives you part of your vertical support. You have no spanner beams in the RCC panels anymore; there goes some of your vertical support. You have no trusses and ribs left inside the structure. Later on, you're probably hot enough in the wing that they may literally be laying in pools on the bottom of the wing. So now the only thing you have keeping the top and bottom part of the wing apart is the RCC panels themselves."

In Mission Control, the first clear sign of a problem aboard Columbia was the loss of data from sensors in the left wing's hydraulic system. The wires leading to those sensors had been part of a cable bundle attached to the outboard wall of the left landing gear wheel well.

As Columbia was crossing the border between California and Nevada, the shuttle's attitude was down to 43.1 miles. But its velocity was still a blistering 22.5 times the speed of sound. It was 8:54:25 a.m.

Observers on the ground saw or photographed more than 10 debris-shedding events in the following few moments.

By 8:56:16 a.m., just four minutes after the forward spar had been breached, the plume inside the wing had burned through the landing gear wheel well box and had begun scorching the massive left main landing gear strut. But the shuttle's computers continued dutifully guiding Columbia toward Florida. Right on schedule, they ordered the nose to swing from right to left in the first of three roll reversals. The shuttle crossed the border between Arizona and New Mexico as the 15-second maneuver was being completed.

Now Columbia's left wing was tilted down toward the ground. The shuttle's flight computers were constantly comparing the ship's actual trajectory with the flight path needed to achieve a landing at Kennedy, adjusting Columbia's elevons to counteract the steadily increasing left-wing drag and the ship's tendency to roll to the right.

As the shuttle streaked across New Mexico, two sensors detected slightly increased pressure in the left main landing gear outboard tire.

Extreme heating would have been required to raise the pressure in the massive, multiple-ply tires and the slight upward trend measured by the sensors hinted at the inferno blazing in the wheel well.

Less than a minute later, at 8:58:03 a.m., Columbia's flight computers detected a sharp change in the aerodynamic forces acting on the shuttle as the depression in the lower surface of the left wing presumably increased in size. At the same time, the drag acting to pull the nose farther to the left continued to increase. Approaching the Texas border, the flight computers again ordered the elevons to counteract the unwanted forces. Several debris-shedding events, indicating the wing was losing additional insulation and structure, were noticed by ground observers.

"The vehicle was in control and was responding to commands up to that point and after that point something changed, apparently," United Space Alliance manager Doug White said at a CAIB hearing. "[The flight system] still continued to be in control and still continued to respond to commands, but the rates and the amount of muscle it needed to continue flying the vehicle the way it should be flown were continuing to increase. Something definitely happened at that point—again, we don't know what—but something definitely happened at that point to cause the flight control system to need more muscle and start having to fight harder to control the vehicle."

Months later, Goodman speculated the sudden change in the shuttle's flying characteristics was caused by a major change in the wing's shape.

"I believe you can make a case. . . . that the wing begins to collapse," Goodman said. "That leads partly to what happens in the wheel well, and you have basically this deformation of the wing all kind of happening simultaneously."

Goodman asked veteran astronaut John Young, an Apollo moonwalker who commanded the first and ninth shuttle missions, if Husband could have felt any vibration or buffeting as the left wing deteriorated.

"I'm a pilot. I asked John Young this question, and John Young said the left wing could fall off and you might not feel it in the cabin."

Hallock explained, saying, "The wing is there to give you some lift when you eventually get into the atmosphere. You're still not really into the atmosphere [deeply] at this point."

The westernmost debris recovered on the ground, the Littlefield tile, came from the upper surface of the left wing just behind the RCC panel 8–9 area. Normally white, its surface was charred and blackened by exposure to extreme heating. The tile fell from the left wing within a few seconds of the sharp change in the aerodynamic forces acting on the shuttle. Goodman said it may have popped off as the wing collapsed.

The sharp change in aerodynamics suggests the major problem facing Columbia at this moment was the increasing pressure of the atmosphere on the weakened wing structure. "It's a transition from having a major thermal event [to] you now have a major structural event," Goodman said. "If you buy the theory that you've just collapsed the wing . . . that is also the same period where your entry debris guys will tell you that's probably where you release the Littlefield tile."

But the crew still would not have noticed any dramatic change.

"The tendency is for it to yaw left, pitch to stay constant, and to roll right," Goodman said. "It's an odd sensation, [but] the crew would never have noted anything because even though that's the tendency to do it, your elevons are still keeping you right on the money."

They did, however, notice the loss of tire pressure data. The computers triggered an alarm in the cockpit and displayed a message to alert Husband to possible problems with the landing gear. This was the crew's first notification of potential trouble. Husband called Mission Control, presumably to report the message—"And, uh, Hou[ston]"—but his transmission was cut off.

Seconds later, at 8:59:06 a.m., a sensor used during landing to confirm the left main landing gear was down and locked just before touchdown suddenly indicated deployment. Rodney Rocha and other engineers saw that reading and grew alarmed. But investigators would later determine the landing gear had not deployed, blaming the false reading on the continuing damage caused by extreme temperatures and burning in the wheel well.

Meanwhile in Mission Control, down arrows on Jeff Kling's computer screen highlighted the lost tire pressure data.

"We just lost tire pressure on the left outboard and left inboard, both tires," he told Cain. By now, Kling was beginning to suspect Columbia was in deep trouble. There was no common thread in the timing or location of the failing sensors. The only thing that could explain the failures, Kling was thinking, was a hole in the wing.

"I was pretty certain we were having some sort of burn-through," he said later. "To get that many things to go at the same time was just too much of a coincidence to be anything else. So things are really starting to stack up. And even though I don't have a full big picture yet, there are a whole lot of things that are pointing in a particular area and pretty much telling me what you expected not to happen was really happening."

Astronaut Charles Hobaugh, sitting to Cain's immediate right, radioed Columbia to let Husband know the flight control team was aware of the alarm and the lost tire data. He added, "And we did not copy your last" to let Husband know he needed to repeat whatever he had been trying to say earlier.

By now, the drag and roll forces acting on Columbia were beginning to reach the point where the elevons could no longer keep the shuttle properly oriented. In seconds, they would reach the limit of their motion.

Husband, perhaps beginning to realize major problems were developing, heard Hobaugh's call and tried to respond.

"Roger, uh, buh . . ."

It was 8:59:32 a.m. and Columbia was approaching Dallas. Seconds earlier, data from the shuttle suddenly froze on the computer screens in Mission Control. Down arrows or the letter S, for "static," had appeared on the screens, indicating the numbers were no longer being updated. Cain and his flight control team assumed communications had been blocked by the shuttle's tail fin and that contact would be restored in a few moments. They knew they had problems, but no one suspected a catastrophe was seconds away.

As it turned out, data were, in fact, still flowing down from Columbia. The signals were garbled, however, and the computers in Mission Control were programmed not to display potentially corrupted information. Investigators later would be able to extract some of the data. That information, combined with readings stored in the MADS recorder, and analysis of recovered wreckage, eventually allowed investigators to develop a rough timeline of events stretching another one minute and 50 seconds beyond Husband's final transmission.

For the astronauts, the final sequence was mercifully brief, but no doubt terrifying.

The left wing had suffered so much damage by now that nothing could be done to keep the nose pointed in the right direction. The elevons were ordered to their limits. It was not enough. First two and then four right-side rocket thrusters were automatically commanded to fire in a futile bid to offset the forces pulling the nose to the left. A master alarm sounded in the cockpit as the elevon control circuitry failed. Columbia's nose yawed farther to the left, toward Earth, as the spacecraft began rolling to its right.

In all likelihood, all or part of the presumably collapsed wing suddenly folded over and broke off. At 8:59:46 a.m., a large piece of debris was seen separating from the shuttle. Labeled "Debris A," this may have been much of the left wing. At almost that same instant, a "roll reference" fault message was generated by the flight computers, indicating a sudden, dramatic change in the shuttle's overall lift and drag.

"What we are postulating is that the nose [swung down] at this point," Cain said later, describing the entry scenario. "At some point right in here the vehicle would break apart. The wing would break off because we think it was the weakest part of the structure."

Columbia's backup flight system computer began generating a string of fault messages. The main computers detected apparent leaks in the left rear rocket pod. Two more large pieces of debris fell away from the shuttle within two seconds of each other starting at 9:00:01 a.m. One of these may have been the shuttle's vertical tail fin ripping off in the hypersonic airstream. The other could have been a large piece of the left-side rocket pod. No one knows.

"Everything just wants to fall over at that point," Cain said. "Because again, this is just like a barn door in wind. If that wing came off as we were falling—pitching down and falling over . . . it is likely that the vehicle then probably broke apart in mid-body area."

But not immediately.

At 9:00:02 a.m., two seconds of relatively clean data reached the ground, providing a snapshot of Columbia's condition at that moment. Some of the data are suspect, but engineers believe on the whole, they are accurate.

Readings showed Columbia's three hydraulic power units were still running, along with the ship's three electrical generators. The main engine compartment was intact, and the communications and navigation equipment in the crew module were functioning normally. The shuttle's life support systems were operational. Air pressure was stable, and the temperature was a comfortable 71.6 degrees.

But all three hydraulic power units had lost pressure, and the ship's reservoirs of hydraulic fluid were empty. The fluid presumably was pumped overboard when the left wing failed 16 seconds earlier. The shuttle's cooling system had shut down, and most of the sensors in or on the left rear braking rocket pod were no longer providing data. Multiple alarm messages intended to alert the crew to problems with the rocket pod were in the process of being generated by the computer system. Extreme temperatures were being recorded by sensors on the belly of the orbiter and along the left side of the fuselage. The electrical system was showing signs of multiple shorts and other problems.

The data do not show the shuttle's orientation during those final seconds. But they do show that orientation was changing at more than 20 degrees per second. It may have been faster. That was the maximum the sensor system was designed to record.

As of 9:00:04 a.m., when the final two seconds of telemetry ended, the fuselage was still intact, along with the right wing and the right rear rocket pod. All or part of the left wing was gone. The condition of the vertical tail fin was unknown.

"Based on the debris and the track, I think that the whole wing disintegrated in pieces and we probably ended up with very little of the left wing actually attached before the entire orbiter broke up," Hubbard said. "I think as they lost control and as the wing started to come apart, the sense from the data we have is that the orbiter went into a [tilted] flat spin, possibly even going backwards for some period of time."

Just before telemetry stopped, data from the backup flight system computer indicated one of the two cockpit "joysticks," used to manually fly the spacecraft on final approach to the runway, was moved beyond its normal position. That's one way for a pilot to deactivate the autopilot. But investigators do not believe Husband or McCool was attempting to take over manual control. More likely, one of the pilots inadvertently bumped his hand controller during those horrifying final few seconds. The shuttle's digital autopilot remained engaged through the final loss of signal.

The MADS recorder operated 17 seconds beyond the final two-second stream of telemetry as multiple pieces of debris fell from the crippled, slowly spinning spacecraft. Finally, at 9:00:19 a.m., the recorder lost power, as the fuselage began breaking apart.

The shuttle was 37 miles up and still traveling 18 times the speed of sound.

At such high velocities, "the heating is like, in a sense, cooking an onion," William Ailor, director of the Aerospace Corporation's Center for Orbital and Reentry Debris Studies, later told the accident board. "You basically start from the outside, and then as you heat the pieces up to a point where the materials will fail, that'll expose some new materials, they'll go through the same process, and the object can be broken apart."

Columbia's breakup generally mirrored the way much smaller, more fragile satellites and rocket stages come apart when falling through the atmosphere.

"Basically," Ailor said, "a typical way for things to break up when they re-enter is they will come down through the atmosphere for a certain amount of time. They look absolutely fine—we've seen videos of these things where they just look like spacecraft coming down. All of the sudden, they come apart, and when they come apart, they just disintegrate."

Throughout the course of the investigation, NASA officials refused to discuss what might have happened to the astronauts after the shuttle broke up. But a team of NASA and contractor engineers conducted a "crew survivability" study for the CAIB that was included in an appendix to the panel's final report.

The study concluded the shuttle's heavily reinforced crew module and nose section broke away from the fuselage relatively intact, separating at the bulkhead that marks the dividing line between the cargo bay and the forward fuselage. Challenger's crew module had also broken away in one piece when the shuttle disintegrated during launch 17 years earlier. As with Challenger, the forces acting on Columbia's crew during this period were not violent enough to cause injury, and investigators believe the astronauts probably survived the initial breakup of the orbiter.

Presumably, the cabin maintained pressure. But there was no electrical power. The shuttle's fuel cells were located under the floor of the cargo bay, and even though it wouldn't have helped at this point, Husband had no way to contact Mission Control. The lights were out on the lower deck, and without power, the intercom system no longer worked.

Like Challenger's crew, the Columbia astronauts met their fates alone and the details will never be known. Clark presumably was still videotaping on the flight deck when the alarms began blaring and the shuttle yawed out of control. But the outer portions of the tape—the portions that might have shown at least the initial moments of the shuttle's destruction—were burned away.

To determine what happened after the module ripped away from Columbia's fuselage, investigators analyzed recovered cabin wreckage and calculated the trajectories the debris items must have followed based on weight and other factors. A sophisticated computer program then used those data to run the trajectories back in time to the point where they intersected, the point where the cabin must have started breaking apart.

On the basis of those data, investigators concluded the module fell intact for 38 seconds after main vehicle breakup, plunging 60,000 feet to an altitude of 26 miles before it began to disintegrate from the combined effects of aerodynamic stress and extreme temperatures. From the debris analysis, investigators believe the module was probably destroyed over a 24-second period beginning at 9:00:58 a.m. During that period, the module fell another 35,000 feet, to an altitude of 19 miles or so.

Investigators believe the module began breaking up at the beginning of that window. If any of the astronauts were still alive at that point, death would have been instantaneous, the result of blunt force trauma, including hypersonic wind blast, and lack of oxygen. Pathologists found no evidence of lethal injuries from heat.

"The most western piece of crew equipment found was a helmet from the mid-deck," according to the CAIB appendix. "The breakdown as to the location of the remaining crew equipment showed that the mid-deck crew equipment was the farthest west and the flight deck crew equipment was at the eastern end of the debris field. Therefore, it seems reasonable to conclude that the crew equipment on the mid-deck separated from the CM [crew module] before the flight deck equipment."

About 45 percent of the crew module was recovered near Hemphill, Texas, including pieces of the forward and aft main bulkheads, the frames from the forward cockpit windows, the crew airlock, and all of the hatches. About three-quarters of the flight deck instrument panels were found, along with 80 percent of the mid-deck floor panels and numerous parts from the crew's seats and attached safety equipment. Wreckage from the ship's galley was recovered, along with parts of the toilet, bailout equipment, tools, one of the crew's sleeping compartments and items that had been stored in middeck lockers. Remarkably, the wristwatch Dave Brown had carried aloft as a belated birthday present for Kennedy engineer Ann Micklos was found, its faceplate missing and its hands frozen at 9:06 A.M.

Condition of the items varied from "highly melted, twisted and torn to near pristine," investigators concluded, noting that crew module debris "experienced noticeably less aerodynamic heating than other portions of the vehicle."

From an analysis of pressure suit components and helmets, investigators concluded three astronauts had not yet donned their gloves when breakup began and one was not wearing his or her helmet. In the end, however, having sealed pressure suits would have made no difference.

But investigators were struck by the way the crew modules of both Challenger and Columbia broke away relatively intact. The survivability study concluded relatively modest design changes might enable future crews to survive long enough to bail out.

But Columbia's crew had no chance. The astronauts fell to Earth amid a cloud of wreckage and debris. One of the crew members came to rest beside a country road near Hemphill. That afternoon, Roger Coday said a prayer and built a cross.

One question that lingered throughout the investigation into the loss of Columbia was what, if anything, NASA could have done to save the astronauts if the severity of the foam strike had been understood early on. It was an emotionally painful issue for many at Johnson because it forced flight controllers, engineers, and managers to consider the possibility that their actions, or lack thereof, had doomed the crew.

Anticipating questions from the CAIB, Cain and a team of NASA and contractor engineers conducted an internal study in March 2003 to determine if the shuttle's trajectory could have been modified enough to significantly reduce heating. As it turned out, the only remotely viable options involved lowering the shuttle's orbit before beginning the descent and staging emergency spacewalks by Brown and Anderson to release the Spacehab module and other payload bay equipment. By drastically cutting the ship's weight by as much as 15 tons, the study concluded, re-entry heating would, in fact, be reduced. But not nearly enough to save the shuttle and its crew.

Cain's team did not look into the possibility of a rescue mission or a spacewalk to plug the hole in the leading edge. But most engineers dismissed those scenarios, believing the former was extremely unrealistic and the latter impossible. Diverting Columbia to the International Space Station was never an option because the two spacecraft were in different orbits.

"I have wracked my brains over this," Cain said at the time. "There just was no way we were getting that vehicle back. If we'd gone and taken some pictures and done whatever else anybody could think of, it wouldn't have changed the outcome for Columbia."

But Gehman and the accident board were not satisfied. Columbia's astronauts had enough food, water, air, fuel, and electrical power to survive a full month—until Feb. 15—if they conserved electricity and were less active. The limiting factor was the amount of lithium hydroxide on board, a chemical used to scrub carbon dioxide from the cabin air supply.

Assuming engineers had realized the severity of the foam strike by the fourth day of the mission, NASA would have had 26 days to either come up with a spacewalk repair plan or to launch a rescue mission. The shuttle Atlantis was scheduled for liftoff on a space station assembly flight just five weeks after Columbia's planned landing. Its boosters and fuel tank were already assembled, and Atlantis was ready for attachment and rollout to the launchpad.

Gehman said board members were disturbed by repeated comments from NASA and contractor engineers and managers that nothing could have been done to save Columbia's crew.

"The implications of the decisions made regarding the photography and the foam strike analysis, and whether or not you should get on-orbit photography, changes from being kind of a bureaucratic, administrative, fumbling, bumbling, to a much more serious life-and-death kind of a decision process," he said in a news conference.

"Because it turns out a lot of people were saying, well, it doesn't make any difference if you take photography or not because there's nothing we can do. Now, those kinds of benign administrative decisions . . . look more ominous because now it looks like maybe there was something you could do. That's the area we were concerned about."

Gehman asked NASA to make two assumptions. The first assumption was that shuttle managers had conclusive proof of a catastrophic breach early in Columbia's mission. The second assumption was that agency managers would commit Atlantis to flight without knowing whether it might fall victim to the same problem that had doomed Columbia.

"This whole question that we asked, and NASA's whole response, is based on a set of assumptions . . . which set conditions which were not present in January of 2003," Gehman told reporters in a teleconference. "In other words, we set a scenario here that was not the scenario of the Columbia. Nobody told us on day 4 that we had a hole in the leading edge of the left wing. So these are two different sets of conditions."

But the bottom line, he said, "is that it's all feasible. There are no show stoppers. But it all turns out to be extraordinarily [difficult]."

Senior flight director John Shannon led the two-week study and reported to Gehman and the board in late May. The team assumed the following: a 6-inch-wide hole on the lower side of RCC panel 8 or a 10-inch piece of missing T-seal between panels 8 and 9; the foam strike was seen on flight day 2; military telescopes or satellites photographed Columbia the next day; and the results were inconclusive. Based on that scenario, the team developed the following timeline:

January 19: Planning begins for a wing inspection spacewalk by Brown and Anderson to resolve the issue once and for all. To protect the option of a rescue mission, engineers at Kennedy are told to begin processing Atlantis for launch, working around the clock in three shifts.
January 20: Brown and Anderson carry out an inspection spacewalk and confirm the presence of a catastrophic breach in the leading edge of the left wing. Johnson engineers begin developing plans for a possible spacewalk repair job to plug the hole. Atlantis processing continues.
January 26: To protect the option of a rescue mission, Atlantis is moved from its hangar to the Vehicle Assembly Building for attachment to its external fuel tank and boosters.
January 30: Atlantis is moved to the launchpad.
February 9: NASA managers decide whether to press ahead with a repair attempt by Columbia's crew or to launch a rescue mission. The earliest possible launch opportunity is Feb. 10.

Shannon's team studied a variety of possible repair techniques, including the feasibility of putting heavy tools in the RCC cavity and then filling a water bag that would freeze in place behind the broken panel. For a missing piece of T-seal, the astronauts could remove tiles from less critical parts of the shuttle, shape them to fit, and then press them into the gap. Two spacewalks would be required: one to scavenge parts and the other to attempt the actual repair.

"They inventoried everything that was on board," Gehman said. "They devised a successful way to get out to the area of the damage without further damage to the TPS [thermal protection system]. They devised a way that they thought they could work out there, and they . . . came up with a patch that they would jam stuff in the hole.

"Then, after a day or two of maneuvering the orbiter so that section of the wing was in the shade all the time, the water would freeze solid, and then that holds the stuff in place," Gehman said. "They would then put something over the top of the hole, some Teflon tape or something like that, and then they would attempt to re-enter."

How long such a patch might hold in the fierce heat of re-entry is not known. But Shannon's team concluded it would not last nearly long enough. In briefing charts presented to the accident board, the repair scenario was described as "very difficult with a low probability of success."

The Atlantis rescue mission on the other hand, was theoretically possible if the key assumptions were met: the damage was unambiguous, and NASA managers were willing to risk another vehicle and crew when the cause of the bipod ramp foam loss in two of the previous three missions was not understood.

To survive that long, Columbia's crew would have had to power down the orbiter and do everything possible to minimize physical activity and their consumption of oxygen. Staging more than one spacewalk probably would have eliminated the Atlantis rescue option because of air lost overboard when the airlock was used.

"The study indicates if you have five or six ifs that line up and you get an affirmative answer to each of the five or six ifs, the launch of the second shuttle and the rescue was conceivable," Gehman said. "It isn't easy; it's not even highly likely. But it is conceivable."

Carrying four veteran astronauts, Atlantis would have had to launch at night to catch up with Columbia within 24 hours. Atlantis would approach from directly below with its payload bay facing Columbia and its left wing pointing in the direction of travel. Columbia would be oriented with its payload bay facing Atlantis, its fuselage at a right angle to the approaching shuttle.

With the spacecraft flying in formation 50 to 60 feet apart, two spacewalkers from Atlantis would carry a pair of spacesuits and additional lithium hydroxide canisters to Columbia. While the first two Columbia astronauts made their way to Atlantis, two others would be donning the fresh spacesuits. The final two astronauts to leave Columbia would set up the shuttle for an unmanned return to Earth on command from Mission Control.

"There was no way to recover Columbia," Gehman said. "The last two astronauts who left the Columbia would flip certain switches in certain positions so Columbia could be deorbited on command from Houston. And then the Columbia would be lost, ditched in the ocean."

All 11 astronauts—four of them sitting on the floor of Atlantis lower deck—then would return to Earth.

Gehman summed up the rescue scenario by saying, "It's technically possible, very, very risky, and a whole bunch of ifs have to line up. . . . But I have no idea if it would have been successful or not."

NASA insiders had little doubt. Few believed the rescue mission was feasible in reality, and many were angered by the discussion.

"We would have been criminally negligent to try to jump into the sky again without knowing what had knocked us out of the sky the last time," one NASA manager said. "Would you really try to hurry up and throw another orbiter and another crew in the sky? As badly as I feel that these people, who I knew personally, had died, could we have done that? Shit no.

"What we would have done was agonized and agonized. We would have stretched out the duration as long as we could, and we would have [had] to deorbit them somewhere and hope that we got at least down to bail-out [altitude]. We would have had to. Think of how much harder that would have been. We would all be in therapy by now."

Even members of the astronaut office at Johnson, who lost friends and colleagues aboard Columbia and who could easily imagine themselves in similar circumstances, were not optimistic about the chances of success. But they wish NASA had at least tried.

"Had we known what we had on orbit, within the first 4 days, I think there was a really good chance we would had gotten the crew back and that is where we naturally excel," said chief astronaut Kent Rominger. "You give us a challenge, we know the problem, and it is amazing what you can do."

Astronaut Jerry Ross also wished NASA could have had a chance, but he was less optimistic about what the outcome would have been.

"I am not sure it would have been feasible, but at least we could have tried," he said. "In some ways, it may be better that we didn't because the crew had a great time all the way up until the very end. Otherwise, they may have been in miserable circumstances for a long time up there and the families would have been hanging on for a long time and it still may have ended up the same way."

Buzzard summed up the feelings of many when he said, "I am not sure we would have committed that next shuttle. That would have been the hardest mission. Two of the last three had foam come off.

"But, the fact that we didn't try made us feel like we let them down."

E L E V E N

RETURNING TO FLIGHT

Based on NASA's history of ignoring external recommendations, or making improvements that atrophy with time, the board has no confidence that the space shuttle can be safely operated for more than a few years based solely on renewed post-accident vigilance.

—*Final report of the Columbia Accident Investigation Board*

With a crackling roar, a Minuteman rocket carrying a dummy warhead thundered away from its silo at the Cape Canaveral Air Force Station on March 14, 1970. It was the intercontinental ballistic missile's final test flight from Florida, and for the next 16 years, the base's two silos sat dormant, buried relics of the Cold War.

The 78-foot-deep launch tubes were re-opened for a final time in late-1986. This time, nothing would be coming out. Instead, something was going in: tons of wreckage from the shuttle Challenger. When the work was completed in January 1987, massive concrete caps were lowered into place, sealing the wreckage in silent darkness.

Today, visitors to this lonely, isolated part of the Air Force station see only the weathered concrete caps. Rusting metal, loose gravel and tall grass give the area a desolate look as ocean breezes swirl across the Cape's long-abandoned complexes 31 and 32. No marker or monument marks this gravesite, where the remains of NASA's second space shuttle are entombed.

The final resting place of the space agency's oldest shuttle would be very different. Eight months after Columbia's destruction in the sky above Texas, Kennedy workers finished moving 42 tons of recovered wreckage out of the hangar where it had been painstakingly reconstructed. Unlike Challenger's remains, debris from Columbia was not sealed off or hidden away. Because it might hold valuable clues for future spacecraft designers, Columbia's wreckage was stored in a cavernous 6,800-square-foot room on the 16th floor of the Vehicle Assembly Building, where the shuttle had begun its trip to the launchpad 10 months earlier.

"We are not burying Columbia," said NASA Deputy Administrator Fred Gregory. "Columbia will be forever used as a learning opportunity, not only for investigators if something similar to this happens, but also as a learning tool for how we better prepare ourselves for our next step."

Surveying the scene from the room's only door, a visitor stares into a rectangular hall 200 feet deep and about 35 feet wide. At the back of the room are spherical tanks that once held liquid hydrogen, oxygen and rocket fuel. Just in front of the tanks are three long rows of stacked 3.5-foot-square boxes holding bar-coded, bubble-wrapped debris. A huge banner hangs from the ceiling just in front of the boxes, signed by the hundreds of engineers and technicians who prepared Columbia for launch and who later helped recover its remains.

Equipment racks line the walls of the room, holding wreckage too large to fit in the boxes. In front of the banner and closer to the door are more recognizable items, including the shuttle's main landing gear, its tires, main engine components and the three-dimensional reconstruction of its left wing leading edge. No longer the center of attention, the fragments of RCC panels 1 through 13 bear mute witness to the fury of re-entry. The MADS recorder is there, along with the window frames from Columbia's cockpit, the shuttle's hatches and the thousands of cards, letters and posters sent in by the public. On the right side of the room are two walled-off areas housing wreckage from Columbia's crew module.

With approval from shuttle program managers, pieces of debris will be retrieved and sent out for detailed analysis to support ongoing investigations into how materials fail in the extreme environment of re-entry. If the wreckage can help designers come up with safer spacecraft, NASA officials believe, the time and effort will be well spent.

Only three people at Kennedy have keys that will unlock the door to the debris room. One of them is Scott Thurston, the engineer who served as Columbia's vehicle manager for its final two missions.

"When you have a chance to sit there quietly and think about it, all the effort and work that went into getting her into orbit and then the human side of it, the crew, it still gets to you," he said. "It still gets to you."

As the Columbia Accident Investigation Board wrapped up its work in late August, NASA managers prepared for the worst.

Many feared the CAIB's recommendations would be so costly and time-consuming that the shuttle would remain grounded indefinitely. The need for shuttle flights to the International Space Station was more urgent than ever. Without shuttle resupply missions, the outpost was basically in survival mode: Operations were severely scaled-back and there was no prospect for finishing assembly until flights missions resumed. In the interim, the project depended on the cash-strapped Russian space program, and NASA was far from certain its former rival could carry the burden alone. That made it crucial in the minds of many NASA officials to get the shuttle flying again as soon as possible.

CAIB chairman Hal Gehman had gone to great lengths to make

sure none of the board's proposed fixes would catch NASA off-guard. All of the report's major themes had been discussed publicly in the board's hearings and press conferences.

"The board decided early on," Gehman said, "that we were going to take a cooperative approach with NASA. The way we liked to describe it was 'This is not a point-scoring contest with NASA to see if we could outscore them.' Indeed, everything that we found or concluded, we wanted them to know about it so they could get to work on getting it fixed."

Investigators often would pass word back to NASA on areas of concern through a pair of space agency officials who were assisting the board—Bryan O'Connor and chief engineer Theron Bradley.

"We would frequently point to them and say, 'Make sure headquarters knows about this,' " Gehman said. "On rare occasions, I would pick up the phone and call [Administrator Sean] O'Keefe, when I thought something really newsworthy was going to come out or something that might be misinterpreted."

During the course of the board's seven-month investigation, the CAIB released four recommendations and publicly discussed other concerns to give NASA a head start. The space agency was listening. By Aug. 5, shuttle managers had put together a 121-page document titled *NASA's Implementation Plan for Return to Flight and Beyond* that was the blueprint for answering the CAIB's anticipated recommendations. To prepare for the report's release, NASA set up an action center to coordinate an agency-wide effort to compare the plan to the CAIB's final recommendations and make revisions where needed. The goal was to have the framework of a final plan for moving ahead in place by early September.

While NASA managers knew the broad outlines of the CAIB report, no one outside the board knew the specifics. The CAIB was obsessed with keeping the report under wraps until its formal release and treated the document like a state secret. Drafts were kept on password-protected government computers that only board members and a few staffers had access to. No one wanted Congress or O'Keefe to read about the findings in the press before they received the report. Many internal NASA reports and a handful of preliminary CAIB documents were, in fact, leaked to the media, but the final report never got out.

At 6 a.m. on Aug. 26, dozens of journalists filed into the National Transportation Safety Board Conference Center at L'Enfant Plaza in Washington for the long-awaited unveiling of what would prove to be one of the most important reports in NASA history.

Reporters were required to check their cell phones before getting a copy; then they were locked incommunicado in the conference center for four hours to review the findings and write their stories. That gave board members a chance to deliver the report to NASA officials and the Columbia astronauts' families before the findings were splashed across television, radio, and the Internet. Congress and the president were out of town that day, but copies were made available to staffers. Gehman and Sheila Widnall personally drove to NASA Headquarters to give a copy to O'Keefe.

Any lingering doubts about the board's independence and the thoroughness of the investigation were quickly erased by the frank, often brutal, tone of the report's opening pages. There were 29 recommendations, 15 of which were expected to be carried out before the shuttle returned to flight. The recommendations ran the gamut from short-term technical fixes to long-term organizational reforms.

It was a report unlike any NASA had seen before:

> The board recognized early on that the accident was probably not an anomalous, random event, but rather likely rooted to some degree in NASA's history and the human space flight program's culture. Accordingly, the board broadened its mandate at the outset to include an investigation of a wide range of historical and organizational issues, including political and budgetary considerations, compromises, and changing priorities over the life of the space shuttle program. The board's conviction regarding the importance of these factors strengthened as the investigation progressed, with the result that this report, in its findings, conclusions, and recommendations, places as much weight on these causal factors as on the more easily understood and corrected physical cause of the accident.
>
> The physical cause of the loss of Columbia and its crew was a breach in the thermal protection system on the leading edge of

> the left wing, caused by a piece of insulating foam which separated from the left bipod ramp section of the external tank at 81.7 seconds after launch, and struck the wing in the vicinity of the lower half of reinforced carbon-carbon panel number 8. During re-entry this breach in the thermal protection system allowed superheated air to penetrate through the leading edge insulation and progressively melt the aluminum structure of the left wing, resulting in a weakening of the structure until increasing aerodynamic forces caused loss of control, failure of the wing, and breakup of the orbiter. This breakup occurred in a flight regime in which, given the current design of the orbiter, there was no possibility for the crew to survive.
>
> The organizational causes of this accident are rooted in the space shuttle program's history and culture, including the original compromises that were required to gain approval for the shuttle, subsequent years of resource constraints, fluctuating priorities, schedule pressures, mischaracterization of the shuttle as operational rather than developmental, and lack of an agreed national vision for human space flight. Cultural traits and organizational practices detrimental to safety were allowed to develop, including: reliance on past success as a substitute for sound engineering practices (such as testing to understand why systems were not performing in accordance with requirements); organizational barriers that prevented effective communication of critical safety information and stifled professional differences of opinion; lack of integrated management across program elements; and the evolution of an informal chain of command and decision-making processes that operated outside the organization's rules.

There were no blockbuster revelations, although many in NASA and the press were surprised by the extent to which the board linked the accident to space station schedule pressures. The real surprise wasn't so much what the report said, as the way it was said. The CAIB minced no words, particularly in the area of NASA's "broken safety culture" and the changes needed to fix it.

"Without these changes," the report said, "we have no confidence that other corrective actions will improve the safety of shuttle operations. The changes we recommend will be difficult to accomplish—and will be internally resisted. . . . The board strongly believes that if these persistent, systemic flaws are not resolved, the scene is set for another accident."

As expected, the report harshly criticized the Mission Management Team at Johnson and its failure to hear the concerns of engineers like Rodney Rocha. The board devoted an entire chapter to flawed decision making at NASA.

"The same NASA whose engineers showed initiative and a solid working knowledge of how to get things done fast had a managerial culture with an allegiance to bureaucracy and cost-efficiency that squelched the engineers' efforts," the report said. "When it came to managers' own actions, however, a different set of rules prevailed. The board found that Mission Management Team decision-making operated outside the rules even as it held its engineers to a stifling protocol."

Those themes were echoed throughout the board's final press conference at 11 a.m. that day. The CAIB had found a pattern of NASA ignoring similar warnings in the past. Investigators were determined not to let it happen again. The report's tough talk was a way to get people's attention.

"NASA management over the years," Gehman explained to reporters, "due to external influences as well as internal influences, has morphed its management structure to where so much authority and power and so much responsibility have been put into one vertical chain, the program manager, that they've lost all of their checks and balances and independent research and independent engineering and the like. That's one problem. The second problem is NASA has been told this ten times. So, they're guilty of two things, and we put that [the report's tough language] in there for emphasis."

There had been uncertainty throughout the investigation as to how emphatically the board would conclude the foam impact was the cause of the accident. After all, no one had actually seen the foam hit the leading edge or taken pictures of damage to Columbia in orbit. The impact tests, however, removed any ambiguity from the minds of investigators.

"In four simple words, the foam did it," Scott Hubbard said, repeating his earlier announcement at the Southwest Research Institute on July 7. "I'll point out one thing about the statement, which is that we do not include the words, 'probably,' 'likely,' 'most probable.' All of this exhaustive work that we've done, all the discussion and the testing, have led us to a simple statement that the foam resulted in the breach that led to the loss of the orbiter."

While the report's technical findings were relatively straightforward, the cultural findings would be considerably more difficult to address. After the Challenger accident, NASA had carefully made organizational changes, but it had slowly drifted back to many of the same unsafe practices. Board members said one of their greatest fears was that the space agency would eventually go down the same road again.

"The board," Gehman said, "feels that there will be so much vigilance and so much zeal and so much attention to detail for the next half-dozen flights, that anything we say probably is an understatement compared to the energy and the diligence that NASA will naturally put into making the first couple of flights safe.

"The natural tendency of all bureaucracies, not just NASA, to morph and migrate away from that diligent attitude is a great concern to the board, because the history of NASA indicates that they've done it before. Therefore, we have a group of recommendations that are designed to prevent that backsliding, or atrophy of energy and zeal, and those are the second group of recommendations that we call 'Continuing to Fly.' And those are more fundamental and harder to do, but they are just as important, perhaps more important than the 'Return-to-Flight' recommendations."

While only six of the board's recommendations dealt specifically with cultural issues, several were sweeping in scope. The CAIB directed NASA to make three of the changes before returning to flight: adopt a realistic shuttle launch schedule "consistent with available resources;" better train Mission Management Team members to deal with possible in-flight emergencies like Columbia's; and draft a detailed plan for creating a new technical engineering authority and reorganizing the shuttle integration office.

The new engineering authority would have sweeping safety responsibilities, including oversight of all technical requirements, sole authority to grant waivers, control over hazard reporting, and independent verification of launch readiness. The final long-term recommendation was to give the safety office at NASA Headquarters more direct authority over the entire safety organization and supply the office with an independent source of funding.

The changes were designed to make the safety organization less beholden to the programs and projects it was supposed to monitor. To truly work, however, the offices would have to function in a more integrated fashion and NASA Headquarters would have to become more involved.

"NASA had conflicting goals of costs, schedule, and safety, and unfortunately, safety lost out in a lot of areas to the mandates of operational requirements," Gen. John Barry, a board member, said. "So, what we went through in our analyses is trying to figure out how we can fix the culture, and it's not an easy task. In order to do that, you have to do some organizational changes."

One board member, Gen. Duane Deal, had grown so concerned with safety practices in shuttle processing at Kennedy that he added an 11-page supplement to the board's report outlining widespread problems. The findings were alarming: Quality control inspectors were unresponsive to workers' concerns. Some inspectors were hired with little or no aerospace experience. Kennedy's quality-assurance workers had lower job classifications than those at other NASA centers. Staffing levels were low. There was little formal training. Too much oversight had been eliminated or shifted from government to contractors. Surveillance and unscheduled spot checks of contractors' work happened rarely, if ever. Inspections usually were conducted only with advance notice.

"Kennedy quality assurance specialist surveillance is essentially nonexistent," Deal's supplement said, because the Kennedy quality program manager discouraged unscheduled inspections. "Testimony revealed that the manager actually threatened those who had conducted such activity, even after a quality assurance specialist had found equipment marked 'ground test only' installed on an orbiter."

Deal insisted the supplement was not a minority report, but an attempt to prevent the findings from being buried among other "observations" that would be easier for NASA to ignore. The goal, he said, was to help prevent the potential cause of the next shuttle accident.

The findings in Deal's supplement weren't the only high-profile issues not included in the board's formal recommendations. Conspicuous by their absence were a pair of potential recommendations the board had debated for months.

One was the long-standing issue of whether to equip shuttles with a more capable crew escape system. A rudimentary bailout system was installed after the Challenger disaster, but it requires the shuttle to be in a controlled glide, below an altitude of 32,000 feet, for the astronauts to escape. Many people, both inside and outside of NASA, considered the system inadequate. The board decided instead to include an observation that any future manned spacecraft should be designed to take advantage of lessons learned from Columbia and Challenger.

"If you define crew escape as the ability to safely extract the crew from any point in the flight," CAIB member Steve Wallace said, "then, to me, you described another re-entry vehicle inside the orbiter. The trades there are huge. . . . The basic approach of doing everything to get the crew back safely [in the shuttle] is the correct approach rather than to say 'Pull the yellow-striped handle between your legs and boom, you will land safely on Earth.' "

Another would-be recommendation that didn't make the cut was a possible requirement that NASA make changes in the shuttle's re-entry flight paths to avoid flying over populated areas. There was widespread concern in the accident's aftermath that only a miracle had prevented deaths on the ground. However, a subsequent study commissioned by the CAIB concluded the odds of casualties were remote—less than a 25 percent chance of a single serious injury. The board decided to include a general observation that NASA should do further studies on the issue and develop a plan to mitigate risk.

The CAIB also decided not to recommend that a flight-safety recorder similar to ones carried aboard commercial airliners be installed on the shuttle. Investigators concluded it would be tough to develop a new device that could survive an accident like Columbia's. The MADS recorder had survived by luck, not design.

With the report now public, perhaps the biggest question remaining was exactly how to go about reforming NASA's culture. The key, board members agreed, would be leadership. But progress would be difficult to track. A question on everyone's mind was how to measure something as subjective as changes in attitudes and mindsets.

Despite the safety shortcomings and failures cited by the CAIB, shuttle commander Eileen Collins never thought twice about flying NASA's first post-Columbia mission, aboard Atlantis.

"Absolutely not. In fact, I am more committed to flying this mission than I ever would have been," she says. "I am going to be extremely confident because look at all this work that is being done, not just done because of [Columbia], but other things that we think are risky. I am so confident; I am so excited; I want to get our country back flying in space again, so I am not one-blink-of-an-eye worried about safety."

Collins' determined enthusiasm comes across in her firm handshake and ready smile. A former test pilot, Air Force Academy math instructor, and mother of two small children, Collins has logged more than 5,000 hours flying time in 30 different types of aircraft and spent more than 22 days in orbit during three shuttle missions.

Collins was NASA's first female shuttle pilot and the first woman to serve as commander, carrying a $1.6 billion X-ray telescope into orbit aboard Columbia in July 1999. It was during that launch that Columbia suffered a fuel leak and a short circuit that left Collins and her crewmates one failure away from a potential disaster.

The soft-spoken Air Force colonel was not the world's first female space commander. That distinction went to Soviet cosmonaut Valentina Tereshkova, who flew aboard a single-seat Vostok 6 capsule in 1963. But Tereshkova was little more than a passenger. As commander of the first post-Columbia mission, Collins would be responsible for a $3 billion orbiter, the lives of six crewmates and, to a very real extent, NASA's future.

"I am a huge believer in the space program, and I want to make sure that we do this right, not just because I want to keep me and my friends safe, but I want our space program to succeed and go on," she said. "For [Columbia's] crew, we want something good to come out of this tragedy."

Born in Elmira, New York, Collins excelled as a student, eventually completing master's degrees in operations research and space systems management. She earned her wings in 1979 at Vance Air Force Base, Okla., and flew as an instructor pilot for the next three years. She then moved on to huge C-141 transport jets at Travis Air Force Base in California.

In 1990, while attending Air Force Test Pilot School at Edwards Air Force Base, Collins was selected by NASA to become an astronaut.

The day Columbia crashed, she was five weeks from launch aboard Atlantis on a space station resupply mission. The goals of the flight were to deliver a fresh three-man crew to the outpost, to bring the lab's outgoing crew home and to ferry up tons of supplies and equipment. Three spacewalks were planned to install a replacement gyroscope and other equipment. Training was at a fever pitch.

But that Saturday morning, the 46-year-old astronaut was at home near Johnson with her 2-year-old son. Her husband and 7-year-old daughter were on a camping trip north of Houston.

"I woke my little boy at 7:30 [central time], I kept saying 'Hey, the space shuttle is going to land' and la-da-da, and I got him all excited and got a little space shuttle model out," she recalled. "I was watching it on TV, and he is asking me, 'Shuttle, shuttle?' "

Then she heard Mission Control report the loss of radar tracking. She instantly knew what that meant. Stunned, she put her little boy "over to the side when it became clear to me that we had an accident . . . I put him at the sink and turned the water on, and that kept him busy for over an hour, if you can believe that."

Within seconds, Collins' thoughts turned to Husband and his crewmates.

"I remember thinking, 'God, why did you do this? Why did you do this to Rick?' He was such a religious person. But then, as the months went by and I thought about this, if anybody was ready, Rick was ready."

Collins called her parents, her sister, and her brother, telling them to turn on their televisions. Stating the obvious, she told them: "We are not going to launch next month."

Shuttle flights were put on indefinite hold, and her crew was split up. Two of them—Ed Lu and cosmonaut Yuri Malenchenko—were launched to the space station the following April aboard a Russian Soyuz rocket, replacing the three-man crew on board at the time of the Columbia mishap. Without periodic shuttle resupply missions, the station could not support three astronauts. The third station flier on Collins' crew, cosmonaut Alexander Kaleri, was launched to the lab complex six months later aboard another Soyuz. He and NASA astronaut Mike Foale replaced Lu and Malenchenko.

In the meantime, Collins and her three other shuttle crewmates—pilot Jim Kelly, Steve Robinson, and Japanese astronaut Soichi Noguchi—could do little more than continue basic training while awaiting the release of the CAIB report.

Because of the step-by-step nature of space station assembly flights, all four would still have seats aboard Atlantis for the first post-Columbia mission. Robinson and Noguchi already were trained for a complex spacewalk to install the replacement gyroscope. But the flight would be very different from the one they originally envisioned. It would no longer be another routine visit to the space station. Atlantis' mission would become symbolic of NASA's response to the CAIB report, a test flight to demonstrate that the technical fixes ordered by the board had, in fact, been implemented.

First and foremost on the list of recommendations was eliminating the foam shedding from the external tank.

The board recommended NASA begin an "aggressive" program to eliminate major foam shedding "with particular emphasis on the region where the bipod struts attach to the external tank."

That one was relatively easy to address. NASA engineers decided to simply remove the ramps and to install electric heaters to prevent ice buildups. The massive fittings that anchor the two struts holding the nose of the shuttle to the external tank—fittings that used to be buried inside the foam ramps—will be fully exposed on all future flights. It was such a simple fix, many wondered why the tank had not been designed that way in the first place.

NASA was told to conduct extensive testing to determine the strength of the wing leading edge panels and the effects of debris strikes, and if possible, to make them less susceptible to impact damage. NASA was also asked to develop reliable methods for inspecting

RCC panels between flights to look for subtle signs of age or weakness that might make a panel prone to failure.

But the most challenging recommendation called for developing ways to spot—and fix—tile and RCC damage in orbit.

"For missions to the International Space Station," the CAIB wrote, "develop a practicable capability to inspect and effect emergency repairs to the widest possible range of damage to the thermal protection system, including both tile and reinforced carbon-carbon, taking advantage of the additional capabilities available when near to or docked at the International Space Station."

All but one of the missions remaining on NASA's launch manifest for the foreseeable future were bound for the International Space Station. For near-term flights, the lab complex would serve as a sort of orbital service station if problems developed, providing additional camera views and a long robot arm to assist spacewalking repair crews if damage was found.

But what about a final flight to service the Hubble Space Telescope, or any future mission that doesn't go to the station? For those cases, the board wrote, NASA must "develop a comprehensive autonomous (independent of station) inspection and repair capability to cover the widest possible range of damage scenarios."

For all missions, regardless of the objective, the board told NASA to create new techniques for the astronauts to carry out detailed inspections on their own, without help from station crews, spy satellites or ground telescopes.

Finally, the board concluded "the ultimate objective should be a fully autonomous capability for all missions to address the possibility that an International Space Station mission fails to achieve the correct orbit, or is damaged during or after undocking."

One glaring issue surrounding Columbia's launch was the lack of clear images showing exactly where the foam struck and the extent to which the left wing was damaged. The one camera that might have shown the impact point was out of focus. The CAIB ordered NASA to upgrade its network of ground-based cameras to provide views of all areas of the shuttle "from liftoff to at least solid rocket booster separation." Making it clear blurry pictures were not acceptable, the CAIB said a fully operational camera system should be a requirement for proceeding with a countdown.

The board also recommended installation of cameras on the shuttle itself and urged NASA to consider mounting cameras on ships or aircraft in the launch area.

For at least the first few flights, NASA officials decided to restrict shuttle launches to daylight hours to provide the best possible lighting for camera views during ascent. In addition, they decided to schedule launches to make sure external tank separation half a world away also occurred in sunlight to guarantee clear pictures of the tank by the astronauts and new cameras mounted on the spacecraft. Those two requirements, along with the nature of the space station's orbit, put severe restrictions on NASA's launch options. In some cases, only a few days would be available each month.

NASA planned to station dozens of state-of-the-art film and video cameras around the launch site to provide overlapping views of shuttles as they climb toward orbit, starting with Atlantis' flight. Cameras bolted to the boosters and tank were intended to provide views of the ship's belly and at least portions of the wing leading edges to spot any debris strikes that might be missed from the ground.

Once in orbit, military telescopes and spy satellite would take pictures of the shuttle. The photos would be passed on to a limited number of NASA engineers with the proper security clearances. The day after launch, all future shuttle crews will carry out their own inspection using a television camera and a laser scanner mounted on the end of a 50-foot long boom. That boom, attached to the end of the shuttle's 50-foot-long robot arm, will enable orbiting crews to inspect all but the rear-most areas of the spacecraft.

But even that was not considered enough. Piloting techniques were developed to manually flip the shuttle 360 degrees as it approaches the space station from below, giving crews aboard the lab complex a chance to photograph the underside of the orbiter from a vantage point just 600 feet away. Collins began practicing the maneuver in shuttle simulators at Johnson within four months of the Columbia disaster.

NASA and contractor engineers carried out tests to determine the level of damage to the tiles and RCC panels that a shuttle could safely re-enter with. Damage larger than that minimum threshold would have to be repaired in orbit.

To fill tile gouges or gaps left by missing tiles, engineers at Johnson adopted an idea developed in the late 1970s but never pursued. With a sort of high-tech caulk gun, a spacewalking astronaut would inject a rubbery, red, silicon-based material called MA-25S into a gouge or gap, using trowels and clear plastic sheeting to fashion a smooth surface. After "curing" in the vacuum of space, the material would form a tough, heat-resistant patch.

Engineers had to figure out how to stabilize a weightless spacewalker, providing support for the astronaut to push against when injecting and shaping tile patches. The solution was to use the space station. Techniques were developed to reposition the orbiter so a spacewalker, anchored to the end of the station's robot arm, could be maneuvered to the damage site.

But techniques for repairing cracked or broken RCC panels posed a tougher engineering challenge. Normal adhesives break down at much lower temperatures than those experienced by the leading edge panels during re-entry. Simply gluing a strip, or panel, of carbon composite material in place over a crack or breach would not work. One promising technique involved filling the cavity behind a ruptured RCC panel with a heat-resistant material that would harden in place like the tile gouge putty. But NASA had no immediate plans for repairing a large leading edge breach like the one responsible for Columbia's destruction. By eliminating the bipod ramps, engineers believed they had removed the only source of potential foam debris large enough to cause such severe damage.

In case any such damage were to occur, however, NASA developed yet another contingency plan: using the international space station as a "safe haven." In this scenario, the astronauts aboard a heavily damaged shuttle would move into the space station and await rescue by another shuttle crew. The damaged orbiter would be set up for an unmanned re-entry and undocked remotely. From there, Mission Control would send commands to scuttle the ship in the Pacific Ocean.

For Collins' crew, the CAIB report triggered a major shift in mission priorities. The robot arm inspections would eat up valuable crew time and add weight to an already fully loaded flight. A spacewalk would be needed to test whatever tile and RCC repair techniques were developed, and again, the required equipment would add unwanted weight. Without a third astronaut aboard the space station, the combined crews would not have time to move all the originally manifested supplies from Atlantis to the lab complex.

In the end, NASA managers decided to split the original mission, known as STS-114, into two flights. The first, with Collins, Kelly, Robinson, Noguchi and three new crew members—Wendy Lawrence, Andy Thomas and Charlie Camarda—retained the original mission designation and at least part of the original payload, including the new gyroscope. Robinson and Noguchi were trained for a tile repair demonstration spacewalk and one of the originally planned excursions was deleted. A second flight with a different crew was added to complete the original resupply objectives of the STS-114 mission.

Rick Hauck, commander of the first post-Challenger mission in 1988, said before launch that his flight would be the safest in shuttle history because of all the scrutiny that went into it. Kelly felt the same way about Atlantis' mission, but added a word of caution.

"For our flight, I think we are going to be the safest that you could possibly be," he said. "But five to 10 years down the road, if we don't get the systemic change that needs to happen for us to see those warning signs or get the checks and balances in place . . . then I don't have much confidence.

"Flying in space is a risky business and sometime in the future we are going to lose another ship and another crew," Kelly added. "My one hope is that when the next accident investigation board concludes its work, they say, 'Well, this was an act of God; it was one of those one-in-a-million chances that is just part of doing business.' And that has got to be the goal, to be as rigorous as we possibly can with enough checks and balances in place . . . that we get to the point where the next accident has nothing to do with the people or the equipment."

In December 2002, just six weeks before Columbia blasted off on its 28th and final voyage, Atlantis commander Collins decided it was time to have a talk with her oldest daughter. With her own launch scheduled for the first week in March, Collins didn't want her to worry.

"She had just turned 7, and I was getting ready to go up in space and I said, OK, this is the time," Collins recalled. "I said, 'Bridget, have you ever heard about the space shuttle Challenger accident?' And she said, 'No, mom.' So I said, 'OK, I am going to tell you about it. I want you to know this from me before you hear it from someone at school.' "

Collins was worried a classmate might say something about Challenger and the fate of the shuttle's crew and she wanted to prepare her daughter, to reassure her nothing like that would ever happen again. "So I told her. I got out a picture of the accident. I got a picture of the crew. I told her what caused the accident," Collins said. "I told her this will never happen again, because it has been fixed."

On the day the CAIB report was released, O'Keefe and Hubbard spoke via satellite to NASA employees around the country. O'Keefe's opening statement quoted remarks made by legendary NASA flight director Gene Kranz in the aftermath of the 1967 Apollo capsule fire that killed three astronauts. The reason quickly became obvious:

> Space flight will never tolerate carelessness, incapacity and neglect. Somewhere, somehow, we screwed up. It could have been a design, in building, or in testing, but whatever it was, we should have caught it. We were too gung ho about the schedule, and we locked out all of the problems we saw each day in our work. Every element of the program was in trouble and so were we. The simulators were not working, Mission Control was behind in virtually every area, and the flight and test procedures changed daily. Nothing we did had any shelf life. Not one of us stood up and said, "Damn it, stop." I don't know what the Thompson Committee [that investigated the Apollo fire] will find as the cause, but I know what I find. We are the cause. We were not ready. We did not do our job. We were rolling the dice.

O'Keefe's message to the NASA rank and file was one of reassurance. Just as NASA had rebounded from fatal accidents in 1967 and 1986, the space agency would safely return to flight and make the fixes the board had recommended. He stressed that senior NASA managers had heard the CAIB's message loud and clear.

But there was still debate over how some of the recommendations would be implemented. The board had told NASA in broad terms what it needed to do to put its safety house in order and fix the shuttle. But it had left the details to the space agency. That made some in Washington more than a little worried. In the past, NASA frequently had paid lip service to reports' findings before going back to its old habits. There also was growing concern the space agency would try to make changes on the cheap without asking for the additional resources needed to do the job right.

O'Keefe however, had consistently pledged to fully comply with both the spirit and the letter of the board's recommendations. He renewed that pledge hours after the report's release but made it clear the agency would do so on its own terms.

"The issue of how we go about picking the options to comply with these recommendations, each one of them, is going to be our charge from this point forward," O'Keefe said. "Admiral Gehman offered the view that one of the other reasons why the recommendations are not directive in terms of how we actually go about compliance is that every member of the board, with all deference to our friend Scott and all of his colleagues, all have a favorite way they'd like to have seen it implemented, for which there was no consensus. . . . So instead, that's our responsibility. We have to implement it. We have to own it."

To help remove doubts about NASA's willingness to follow the recommendations, the agency chartered a panel chaired by former Apollo astronaut Tom Stafford and Dick Covey, a former shuttle commander, to assess the agency's response. The Return to Flight Task Group planned to submit its final report one month before the shuttle's first post-accident launch.

The task group included numerous former NASA managers, including Jim Adamson, an aerospace executive and former astronaut. He said the key element of the panel's charter would be assessing NASA's intentions.

"Our job, on the surface, may sound simple, but we're going to check and see if they've done that," Adamson said. "It's not a simple yes-or-no question. I think everybody knows there's more than one way to skin a cat, and we're going to try to look, to drill down into NASA's response to make sure they've met the intent of the recommendation."

Covey, who copiloted the first post-Challenger mission, cautioned reporters that he did not expect the panel to address the broader cultural issues cited by the CAIB. Those recommendations were not designated "return-to-flight" items and as such, "the real implementation may take longer," he said. His response again raised the issue of who, if anyone, would make sure NASA had changed its ways organizationally.

At an Aug. 27 news conference, O'Keefe insisted agency managers understood the need for a new way of thinking.

"We get it. We clearly got the point," he said. "There is just no question that is one of their [the CAIB's] primary observations, that what we need to do and what we need to be focused on is to examine those cultural procedures, those systems, the way we do business, the principles and the values that we adhere to."

For O'Keefe, there probably was a sense of déjà vu as he stood before the media to assure them NASA understood the gravity of the CAIB's findings. More than a decade earlier, in September 1992, then-Secretary of the Navy O'Keefe had addressed reporters in the Pentagon's briefing room after the release of a tough report on the Tailhook scandal. Dozens of women had been groped and assaulted during a wild 1991 naval aviators' convention in Las Vegas.

"I need to emphasize a very, very important message: 'We get it,' " he told reporters in 1992. "We know that the larger issue is a cultural problem."

O'Keefe had gotten generally high marks for cleaning up the Tailhook mess and getting the department back on track. Eleven years later in the NASA Headquarters auditorium, O'Keefe said he took full responsibility for the CAIB's findings. One reporter asked him if that meant he planned to step down or felt any pressure to resign. O'Keefe replied that he served at the pleasure of the president.

"I will adhere to his judgment always on any matter, including that one," O'Keefe said. "And so no, there is nothing that in my mind transcends that requirement, and I intend to be guided by his judgment in that regard."

In fact, almost no one in Washington held O'Keefe personally responsible for the accident. He had been NASA administrator for slightly more than a year at the time of Columbia's launch. The report made it clear NASA's deep-rooted cultural problems had taken much longer to evolve. There was growing frustration among some in Congress, however, about his continued reluctance to make the sort of personnel changes at NASA Headquarters that he had among shuttle managers at Johnson.

The previous day, O'Keefe had gotten a vote of confidence of sorts in a brief, three-paragraph prepared statement released by the White House. In stark contrast to President Reagan's ceremonial acceptance of the Rogers Commission's report, President Bush made no appearance to speak publicly about the CAIB's findings.

"The next steps for NASA, under Sean O'Keefe's leadership, must be determined after a thorough review of the entire report, including its recommendations," part of the statement said. "And our journey into space will go on. The work of the crew of the Columbia and the heroic explorers who traveled before them will continue."

Even so, some, like CAIB consultant Diane Vaughan, remained skeptical about whether NASA officials had, in fact, "gotten it."

"I don't feel that they get it," Vaughan told the *Orlando Sentinel*. "[O'Keefe] did talk about his personal accountability, and I think that's very important, but I haven't seen any indication that they understand the systemic nature of the problem that goes beyond individual accountability."

The issue of accountability would dog O'Keefe the following week. He and Gehman appeared together before the Senate Commerce, Science and Transportation Committee on Sept. 3 to discuss the CAIB's findings. Gehman told the committee in his opening statement that the board believed in accountability and had included plenty of information in the report on the actions of individuals.

"I have defended the position before this committee before, that we were not going to make those judgments," Gehman said. "But we put it all in the report. It's all there. If somebody, the administrator of NASA or this committee, wants to find out whose performance was not up to standard, it's all in the report."

The committee's chairman, Sen. John McCain of Arizona, used the opening to ask O'Keefe what sort of accountability he intended to enforce. O'Keefe replied that 14 of 15 senior shuttle managers had been given their current assignments in the wake of the disaster. He did not mention, however, that none of the senior human spaceflight or safety officials at headquarters had been removed or reassigned.

McCain asked if the job shifts that had occurred meant those who were replaced were being held accountable.

"Those who are not there," O'Keefe responded, "I think you can draw the conclusion from that."

Although the CAIB had found agency-wide cultural, safety and decision-making problems, only managers at Johnson were specifically disciplined after the disaster. In fact, the public affairs office at Marshall had gone to great lengths a week earlier to knock down a news story that said Jerry Smelser had been "removed" as manager of the External Tank Project instead of transferred to another job at his own request.

O'Keefe had told senators in May there would be "no ambiguity on this question of accountability at all" when the Columbia probe was completed. However, a week after the investigation was finished, the issue was more muddled than ever.

The next questioner, Sen. Ernest Hollings of South Carolina, was out for blood. He asked Gehman if O'Keefe should be held accountable.

"Almost everything we complain about, every management trait, every communications problem, every engineering problem that we complain about in this report, was set in motion between 5 and 15 years ago," Gehman replied. "So it didn't happen on his watch."

Hollings turned next to the cases of Linda Ham and Ralph Roe. He was furious that the pair who had played a role in the ill-fated photo request remained in positions of authority.

"Mr. O'Keefe has made a very categorical and convincing statement about 'We've got the message,' " Hollings said. But the imagery request, the senator pointed out, "was cancelled by none other than

Linda Ham, who has now just been reassigned over in Houston to another office. And of all things, you say, 'I get it. We're going to do it categorically. We're going to take on every issue. We're going to do everything.' We've put Mr. Roe as number two at the safety office. That doesn't indicate to me that you got it."

Roe, in fact, had not been named the safety office's second in command, but Hollings' point was clear, nonetheless: Heads needed to roll—and not just at Johnson. Following the hearing, O'Keefe refused to discuss specific NASA managers by name with reporters, insisting there was no need for a "public execution." He said the personnel moves were finished—with no changes to senior management in Washington.

Months later, O'Keefe would explain his reasoning:

"In terms of the very senior people here at headquarters, Bill Readdy had become the associate administrator [for spaceflight] six months before the accident. Bryan O'Connor had come to the job as the safety and mission assurance associate administrator eight months before the accident. [Mike] Kostelnik had come in and been recruited just that summer [of 2002]. All of these people were in relatively new capacities . . . There was nothing that I could see in terms of their actions that had called in question their direct involvement or judgment—that was so errored as to have been a major contributing factor to what occurred on that terrible day. Did all of us make judgments that would have, in hindsight, been better if we had done something else? I think the facts of that are pretty damn clear."

During the hearing, Gehman noted the shuttle program's spending power had dropped by 40 percent and that underfunded budgets "had a lot to do with what happened." As senators tried to pin down O'Keefe on how much money would be needed for the shuttle and future space goals, the former White House budget deputy steadfastly refused to discuss timetables or additional funding, saying they would be determined by the president.

"To the extent that a request is required, it will be at the point in which the president determines that that's necessary," O'Keefe said in a brusque exchange with Bill Nelson, a senator from Florida. "And that's exactly when it will be delivered."

Cynics in both parties interpreted O'Keefe's obstinance to mean there would be no major funding increases for the shuttle. Instead, the space agency would likely resort to its old habit of robbing Peter to pay Paul, wringing more money out of other NASA programs. In the months after the accident, it had become clear there was a willingness among many in Congress to support increased spending on a successor spacecraft or shuttle improvements on a scale not seen in years. On Oct. 24, a bipartisan group of 101 concerned House members sent a letter to the White House to "clearly and unambiguously express our strong interest in reinvigorating NASA and turning this funding around." Twenty-three senators sent a similar letter on Nov. 14. It appeared clear, however, that the Bush administration was reluctant to act on those sentiments if that meant spending significantly more money.

Four weeks after the report's release, Bush offered only vague hints about his ideas on the future of space exploration to a small gathering of reporters.

"We've got an interagency study going on now that will enlighten us as to the best recommendations necessary for NASA to proceed in a way that is a good use of taxpayer dollars," Bush said. He added that he didn't know exactly what those goals were yet. "I really don't have an opinion on Mars, but I do have an opinion that the more we explore, the better off America is. . . . I believe in pushing the boundaries."

NASA supporters had heard the phrase "good use of taxpayer dollars" many times before, from Congress as well as the White House. Many considered the expression a Washington code meaning there would be no new initiatives that cost serious money. However, an interagency group, including O'Keefe, Vice President Dick Cheney, and members of the Bush administration's Domestic Policy Council and National Security Council, had been studying ideas for a new space agenda since spring 2003. As the year ended, one of the goals being considered was returning astronauts to the moon to develop technologies for future manned missions to Mars and beyond. But any ambitious new programs would cost lots of money.

The CAIB report had noted many of NASA's problems were rooted in a lack of purpose and clear goals—problems that several presidential administrations and congresses shared the blame for.

"A strong indicator of the priority the national political leadership assigns to a federally funded activity is its budget," the report said. "During the past decade, neither the White House nor Congress has been interested in 'a reinvigorated space program.' Instead, the goal has been a program that would continue to produce valuable scientific and symbolic payoffs for the nation without a need for increased budgets."

The day after Gehman and O'Keefe's joint appearance before the Senate, Gehman and fellow board members Sheila Widnall, Jim Hallock and Maj. Gen. Ken Hess appeared before the House Science Committee. As member after member praised the investigators for their work, Chairman Sherwood Boehlert, a Republican congressman from New York, asked if the CAIB would return in a year to report on NASA's progress in implementing their recommendations. The board agreed. Now, there would be another set of eyes besides the Stafford-Covey group making sure the recommendations were being followed.

Boehlert also mentioned NASA would not get a blank check to return the shuttle to flight or meet its other ambitions.

"When it is pointed out that we don't provide the budget consistent with an agency's ambitions," Boehlert said, "I would suggest the agency better adjust its ambitions."

That afternoon, the Senate Appropriations Committee voted to cut $200 million from NASA's proposed $15.5 billion budget for 2004.

EPILOGUE

Based on its in-depth examination of the space shuttle program, the board has reached an inescapable conclusion: Because of the risks inherent in the original design of the space shuttle, because that design was based in many respects on now-obsolete technologies, and because the shuttle is an aging system but still developmental in character, it is in the nation's interest to replace the shuttle as soon as possible.

—*Final report of the Columbia Accident Investigation Board*

In April 1968, the year before Neil Armstrong and Buzz Aldrin landed on the moon, Stanley Kubrick and Arthur C. Clarke's *2001: A Space Odyssey* imagined the swift development of huge space stations complete with name-brand hotels, airliner-type commercial flights to low-Earth orbit, and sprawling lunar bases serviced by fleets of shuttlecraft. In the heady climate of the Cold War space race, Kubrick's grandiose vision did not seem outlandish. Eight months after the movie's U.S. premiere, Apollo 8 astronauts Frank Borman, Jim Lovell, and Bill Anders rocketed to the moon, becoming the first humans to leave Earth's orbit. Six months later, in July 1969, Armstrong piloted the Apollo 11 lunar lander to a historic touchdown on the Sea of Tranquility.

It took the United States just eight years to go from launching the first American into space to sending astronauts to the surface of the moon. No one knew what the next 30 years might bring but with plans already in the works for proceeding on to Mars by the early 1980s, a real-life space odyssey didn't seem like sheer fantasy to the moviegoers of 1968.

Reality, however, did not keep pace with Hollywood. Three moon missions were cancelled as the nation turned its attention to unrest at home and the war in Vietnam. The agency's budget shrank, and thousands of aerospace engineers were laid off. The Apollo program became a distant memory, a glorious quirk of 20th century international politics. Instead of following Apollo with missions to Mars, NASA was forced to settle for low-Earth orbit. America launched the short-lived Skylab space station while beginning construction of what would become the space shuttle Columbia.

Now, more than three decades after Kubrick's hopeful vision, America's space program has been reduced to an unfinished International Space Station accommodating two inhabitants and a three-shuttle fleet hobbled by age, increasingly tight budgets, and, after two disasters, a crippling loss of confidence.

The space shuttle program has spanned almost a quarter of aviation history in the century since the Wright brothers first flight in 1903. Before Columbia was lost, many in the shuttle program and some in Congress spoke enthusiastically about flying the ships until 2020—a projected lifespan of almost 40 years. In comparison, it took humankind only slightly longer to go from riding horseback to riding rockets into space.

The shuttle remains an international symbol of American technological prowess, despite the fact that much of its design is based on 1970s engineering. Its capabilities are unmatched, even by would-be replacements currently found only on PowerPoint slides. It's also one of the most reliable manned launch systems ever built, the Columbia and Challenger accidents notwithstanding. However, the shuttle has two fatal flaws it can never overcome: It costs too much to fly—a half-billion dollars per mission—and it can never be made dramatically safer, no matter what is done to upgrade its aging systems. NASA estimates the chance of a launch disaster is about 1 in 500 every time the shuttle lifts off. The space agency announced a goal in the late 1990s of raising those odds to 1 in 10,000 for U.S. manned missions, but making the shuttle that safe is simply impossible.

As for flying it for another two decades, the Columbia Accident Investigation Board recommended that NASA be forced to recertify the shuttle system if the nation opts to use it beyond 2010. Recertification would be a costly, time-consuming process to re-verify the engineering assumptions that went into the design in the first place, to recheck the safety margins of its myriad subsystems, and to carry out detailed inspections of every nook and cranny to make sure no age-related problems pose a threat to safe operation. The recertification requirement alone could spell the end of the shuttle program.

For those and other reasons, the Columbia Accident Investigation Board emphatically noted the need to replace the aging fleet as soon as possible. But the same question that has haunted NASA for more than a decade remains: With what?

"We criticize the United States for finding ourselves in a position where we are right now, where we don't even have a vehicle on the drawing board," Hal Gehman said when the board's report was released. "We do have some sense of urgency that another human-carrying vehicle needs to come along fairly quickly."

NASA's efforts along those lines in recent years offer little reason for optimism. After the Clinton White House announced plans to find a shuttle replacement in the mid-1990s, a variety of companies—some big, some small—began to look at ways of putting people into space more cheaply. A projected boom in satellite launches provided commercial incentive, and there was hope among many that industry would build launchers and eventually take over space travel just as commercial aviation had blossomed many years earlier. NASA would simply hire rides on privately owned spaceships. It sounded like a grand idea until reality set in. The potential for profit evaporated almost overnight when the predicted boom in satellite launches never materialized. And development of new launch systems turned out to be even more hugely expensive than expected.

The most spectacular failure was X-33, a joint NASA–Lockheed Martin project to build a ship that could reach orbit in a single step without shedding spent boosters or fuel tanks, launching like a rocket and returning to Earth as an airplane. The program began in 1996, but engineering problems quickly pushed it behind schedule and over budget. After spending $1.3 billion on X-33—$912 million of it taxpayer money—NASA pulled the plug in March 2001 without a single test flight.

A more recent idea for putting Americans in orbit is a small space plane that would lift off atop a commercial expendable rocket, in much the same way astronauts were launched in the 1960s. As currently envisioned, the orbital space plane would principally be used to do little more than ferry crews to and from the International Space Station. Even that modest goal ran into roadblocks over design, cost, and schedule. At the end of 2003, cost estimates remained elusive and a goal of having a space plane operational by 2008 appeared extremely optimistic. Congress was sharply critical, demanding more details and concrete answers on when the vehicles would be ready and what they would be able to do. Lawmakers had ripped X-33 three years earlier for being a so-called single point design, meaning NASA decided on a ship and its capabilities before developing the needed technologies. Now, Congress criticized the space plane program because it was a collection of technology development projects with no agreed-upon vehicle design.

Building a fleet of ships with far fewer capabilities than the shuttle primarily to travel back and forth from the space station also begged more fundamental questions: Does NASA need a small spaceplane or a bonafide shuttle replacement? Can the government afford to replace the shuttle? Does America really need the station? Is it worth the colossal investment needed to build and operate it—an investment that would span more than two decades—or is it a monumental boondoggle?

Viewed on technical merits alone, the International Space Station is an impressive machine with capabilities that dwarf those of NASA's three-man Skylab and any of the orbital labs launched by the former Soviet Union. When complete, the international station's huge solar arrays will generate enough power to run life-support systems, computer networks, and state-of-the-art experiments in a chain of linked-together U.S., Russian, Japanese, and European research modules.

While a far cry from the great wheel-shaped station in Kubrick's *Space Odyssey*, the completed international complex will be bigger than two football fields side by side, weigh more than 1 million pounds, and have the living space of a five-bedroom house. Moving through space at 5 miles per second, it will outshine everything in the night sky except the moon and Venus.

But the station has never ignited the public's imagination—or Washington's—the way the Apollo program once did.

Supporters have long argued the high cost of the station is justified by a wide variety of factors, not all of them directly related to spaceflight. Scientific research is the overarching rationale, but others include international cooperation, the involvement of Russian rocket scientists in a peaceful endeavor, and the ongoing development of U.S. aerospace technology. To many true believers, complex justifications don't really matter. It's enough to simply establish a permanent human presence in space. Others believe a station of some sort is needed to collect medical data on the long-term effects of weightlessness before astronauts can venture on to Mars or beyond. Some support the current station out of fear that its demise could end the American manned space program for the foreseeable future. For them, it's the only game in town.

Critics argue such thinking is shortsighted and counterproductive. They say the station amounts to little more than a taxpayer-funded jobs program for huge aerospace companies and that the research currently planned will never justify the price tag. While few argue with the benefits of international cooperation—any future Mars mission almost certainly will be an international project—the critics say the station's enormous cost outweighs any potential benefits. The station, and the shuttle operations required to support it, total almost a third of NASA's $15 billion annual budget. They are such a huge drain that it's virtually impossible for the space agency to tackle other manned programs without killing the outpost or getting a major funding increase.

When Sean O'Keefe took over as NASA administrator, the station program was projected to be nearly $5 billion over budget. O'Keefe promptly put the design on hold, killed plans for a sophisticated emergency escape craft and eliminated two U.S. modules. He said NASA would complete installation of the station's solar arrays and a multihatch module that would enable the eventual attachment of European and Japanese research modules. The crew size would be frozen at three in this "core complete" configuration while NASA evaluated lower-cost alternatives for implementing the original design. The "schedule pressure" cited by the CAIB as a factor in the Columbia disaster was driven by the directive from NASA Headquarters to achieve the core complete configuration on time and on budget. It's now more unclear than ever exactly when the outpost will be completed and what it ultimately will look like. About $25 billion has already been spent on U.S. components. Including launch costs, operating expenses, and the contributions of 15 international partners, the total price tag is expected to top $100 billion.

Like the station's future, America's destiny in space is inextricably tied to Congress' willingness to pay the bills. But without a clear vision, getting more money for NASA is an increasingly tough sell.

At the dawn of the Space Age in the early 1960s, NASA's mission was crystal clear: Beat the Soviet Union to the moon, and show the world that American technical prowess and free enterprise were more than a match for communism and a planned economy. NASA accomplished the feat in spectacular fashion, with Washington spending nearly four cents of every tax dollar on the space program with the full support of the American public. Today, NASA spends considerably less than a penny of every tax dollar, and there is little reason to think that will change anytime soon.

Gehman and other CAIB members said on many occasions they hoped their report would spark a national debate on America's future in space. By the fall of 2003, that debate was starting to materialize. The Bush administration floated what appeared to be several trial balloons gauging public interest in a return to the moon by 2020. The idea would be to develop technologies in the process that could eventually be used to travel to Mars and beyond. While the initial response was generally favorable, the ever-present question of how to pay for such a costly new initiative remained. As NASA neared the first anniversary of the Columbia accident, it wasn't clear the space agency would even get an additional $280 million believed necessary to return the shuttle fleet to flight.

One Democratic congressional staffer summed up the view of many in Washington on the Bush White House's civil space plans: "Talk is cheap. When they put real money where their mouths are, then maybe I'll believe."

When asked separately in recent months what NASA's central mission should be, the agency's last two administrators, Dan Goldin and O'Keefe, answered with identical one-word responses: exploration. However, critics argue that hundreds of trips into low-Earth orbit for scientific experiments of debatable value is hardly genuine exploration.

A growing number of researchers insist that if the goal is pure science, unmanned missions reap greater rewards at a fraction of the cost with little or no risk to human life.

In the past four decades, robotic probes have studied the sun in unprecedented detail and visited every planet in the solar system except remote Pluto. Comets have been photographed at close range, and landers have touched down on Venus, the moon and Mars. Orbiters have circled Venus, Mars and Jupiter for long-term observations. One even landed on an asteroid at the end of its mission. The Voyager probes, after flying past Jupiter, Saturn, Uranus and Neptune, are still beaming back data as they search for the boundary between the solar system and interstellar space. Robotic spacecraft now on the drawing boards may one day find Earth-like planets orbiting distant suns, perhaps even detecting the chemical byproducts of life. Looking farther afield, unmanned observatories like the Hubble Space Telescope have revolutionized humanity's understanding of the birth, evolution and eventual fate of the universe, shedding light on galaxy formation, black holes and other exotic phenomena.

Nearly a dozen of those spacecraft were launched from space shuttles and one of them—Hubble—was repaired by spacewalking astronauts. But no more. Even advocates of manned spaceflight now agree it makes more sense to use expendable rockets for such missions of exploration. Critics go further, contending humans should be taken out of the loop completely. They believe unmanned missions can outperform anything astronauts can hope to accomplish on the shuttle or space station, with the exceptions of medical and biological research. And those objectives, critics say, could be accomplished with a less-costly outpost. As the late astronomer Carl Sagan once noted, the moment you put people aboard any space mission, the top priority changes from doing science to returning those people alive.

But NASA continues to "sell" its human space program as a vehicle for both science and exploration.

"It fits a romantic model we have of old-fashioned heroic explorers and pioneers," Duke University historian Alex Roland, a long-time NASA critic, told the *Orlando Sentinel*. "NASA has done everything in its power to perpetuate that romantic myth and turn manned space-flight not into useful work but into public spectacle."

But the manned space program remains an integral part of America's self-image. Even though pictures of the astronauts riding moon buggies on the lunar surface don't hold a scientific candle to Voyager's unprecedented views of Neptune, for example, they remain a source of national pride. But is it worth the enormous cost?

The International Space Station has utterly failed to capture the public's imagination or, more damningly, that of the scientific community at large. The station stands largely as an end in itself, not as a stepping stone to some loftier goal. In that light, the lab's high cost becomes the only relevant factor in the ongoing debate about the nation's future in space.

If human exploration is deemed a worthwhile reason to sustain massive manned space programs, there are two obvious destinations: the moon and Mars.

In a 2001 interview, lunar scientist Alan Binder argued the moon would provide an ideal location for low-gravity research and exploration. Even tourism. Raw materials would be readily available for construction projects, and solar power stations could eventually drive industrial development.

"Lunar science never sold itself," Binder said. "We as a community have not made people understand the value of the moon to the Earth in terms of resources and opening up the rest of the solar system. I firmly believe the moon is the key to the rest of the solar system. Once you've got facilities on the moon, industrial facilities, then you've opened up the full-scale, manned exploration of Mars and everything else. You can do it more cheaply."

China's military space officials have made no secret of their long-range desire to send their own astronauts to the moon and to eventually establish a base there. But while rumors of a possible return swirled in 2003, there remained no concrete plans.

"We've been to the moon," Binder laments. "Mars is much more intriguing to the common person because of the possibility of life. It's kind of Earth-like, etc. And since we've already been to the moon, people have kind of gotten over that."

To many, Mars is the next logical step in solar system exploration, a target that would do everything the space station can't in terms of inspiring the nation and providing a grand purpose. It also is the most convenient place to start looking for an answer to one of humankind's most fundamental questions: Has life ever existed anywhere else in the universe?

NASA currently is engaged in a long-term program to explore Mars with automated spacecraft to find out whether the water that once carved its surface remained in place long enough for life to evolve. The long-range goal is to mount a robotic mission to scoop up rock and soil samples and return them to Earth for laboratory analysis.

While NASA planners talk of someday sending astronauts to the Red Planet, there are no such missions currently funded or even on the drawing board. Cost estimates are pure fantasy. But Mars is such a compelling target, thousands of scientists, engineers and enthusiasts are actively lobbying for manned missions, arguing humanity is better prepared for such a flight today than NASA was when it undertook the Apollo moon program.

At the heart of the argument is the possibility of making the sort of scientific discovery that would dwarf anything to come out of the space program to date.

"Our robotic probes have revealed that Mars was once a warm and wet planet, suitable for hosting life's origin," reads the charter of the Mars Society, a grassroots advocacy group. "But did it? A search for fossils on the Martian surface or microbes in groundwater below could provide the answer. If found, they would show that the origin of life is not unique to the Earth, and, by implication, reveal a universe that is filled with life and probably intelligence as well. From the point of view learning our true place in the universe, this would be the most important scientific enlightenment since Copernicus."

Enormous technical issues, including how to keep astronauts healthy for such extended missions remain, along with the question of how much it will cost.

Regardless of the target, NASA must develop a safer, less expensive way to ferry men and women from Earth's surface to low-Earth orbit. It costs roughly $10,000 to launch a pound of cargo—human or otherwise—into orbit aboard the space shuttle. Until launch costs go down, space exploration will remain a government monopoly and the price of doing business will continue to drive the debate.

The lesson of the Columbia disaster is clear to all but the most diehard supporters. The shuttle should be retired as soon as possible. Spending a half-billion dollars a flight because the nation is unwilling to make the up-front investment in a new means of reaching space is penny-wise and pound-foolish. Sooner or later, another mishap is inevitable, and without another option in the wings, America's manned space program will very likely collapse. Many believe the shuttle program survived the Columbia disaster only because of the presence of a crew aboard the International Space Station and the money the project funnels into congressional districts across the United States.

NASA almost certainly will use the space shuttle to complete initial assembly of the space station, if for no other reason than to protect the nation's investment and honor its international commitments. After that, the space agency is looking at operating the shuttle as an unmanned, remotely piloted cargo ship as soon as a possible orbital spaceplane becomes available to ferry crews up to and down from the station. Another possibility is retiring the shuttle as soon as 2010, when recertification would be needed to continue flying.

In the absence of a coherent national space strategy, NASA and the nation will remain mired in low Earth orbit. Without a long-range goal beyond the space station, human missions out into the solar system and beyond will remain the stuff of science fiction, fantastic dreams confined to the imaginations of artists like Kubrick and Clarke instead of real-life aerospace engineers.

The Columbia accident has left NASA at a crossroads. The direction the agency chooses will go a long way toward determining whether NASA—and America—still have the right stuff where space is concerned.

Columbia commander Rick Husband, a devout Christian, undoubtedly was familiar with Solomon's words as recorded in the Old Testament book of Proverbs. The ancient king's wisdom remains as relevant to a space program at the turn of the 21st century as it was to his subjects some 3,000 years ago:

"Where there is no vision, the people perish."

APPENDIX 1

MAJOR PARTICIPANTS IN COLUMBIA'S MISSION AND ACCIDENT INVESTIGATION

Altemus, Steve NASA test director and manager of the Columbia reconstruction effort at Kennedy
Anderson, Michael Columbia astronaut and payload commander
Austin, Lambert a shuttle manager and head of NASA's integration office at Johnson

Barry, John CAIB member and Air Force major general
Benz, Frank NASA's director of engineering at Johnson
Brown, David Columbia astronaut
Buzzard, Frank director of the NASA task force supporting the CAIB

Cabana, Bob astronaut and chief of flight crew operations at Johnson
Cain, LeRoy NASA flight director for Columbia's launch and re-entry
Campbell, Carlisle NASA structural engineer and landing gear expert at Johnson
Card, Mike NASA manager in the headquarters' Safety and Mission Assurance Office
Chawla, Kalpana Columbia astronaut and flight engineer
Clark, Laurel Columbia astronaut

Daugherty, Bob NASA engineer and landing gear expert at Langley Research Center

Deal, Duane CAIB member and Air Force brigadier general

Diggins, Dan CAIB investigator from the FAA

Dittemore, Ron NASA's shuttle program manager at Johnson

Dunham, Mike Boeing engineer on Johnson's debris assessment team

Engelauf, Phil flight director, mission operations representative at Johnson and member of the Mission Management Team

Gehman, Harold CAIB chair and retired Navy admiral

Goldin, Dan former NASA administrator from 1992 to 2001

Gregory, Fred NASA deputy administrator and former safety chief at headquarters

Hale, Wayne NASA flight director and launch integration manager at Kennedy

Hallock, Jim CAIB member and aviation safety expert

Halsell, Jim astronaut and former launch integration manager at Kennedy

Ham, Linda chair of the Mission Management Team

Hartsfield, James NASA public affairs officer at Johnson

Heflin, Milt NASA's chief flight director at Johnson

Hess, Ken CAIB member and Air Force major general

Hill, Paul NASA flight director

Hubbard, Scott CAIB member and director of NASA's Ames Research Center

Husband, Rick Columbia astronaut and mission commander

Kling, Jeff a NASA mechanical systems officer in Mission Control at Johnson

Kostelnik, Mike deputy associate administrator for spaceflight at NASA Headquarters

Leinbach, Mike NASA launch director at Kennedy

Logsdon, John CAIB member and director of the Space Policy Institute at George Washington University

Madera, Pam United Space Alliance engineer and chair of the debris assessment team

McCool, Willie Columbia astronaut and pilot

McCormack, Don a NASA manager of Johnson's Mission Evaluation Room

O'Connor, Bryan associate administrator for safety at NASA Headquarters
O'Keefe, Sean NASA administrator appointed in 2001
Oliu, Armando NASA engineer at Kennedy on the ice, debris and final inspection team
Ortiz, Carlos a Boeing engineer at Johnson on the debris assessment team
Osheroff, Douglas CAIB member and Nobel laureate in physics at Stanford University
Otte, Neil deputy director of NASA's External Tank Project Office at Marshall

Page, Bob NASA engineer at Kennedy and chair of the Intercenter Photo Working Group
Petete, Trish a deputy manager in NASA's Orbiter Project Office at Johnson
Phillips, Dave NASA liaison at Patrick Air Force Base in Florida

Ramon, Ilan Columbia astronaut and first Israeli in space
Readdy, Bill associate administrator for spaceflight at NASA Headquarters
Ride, Sally CAIB member, former astronaut and physics professor at the University of California, San Diego
Rocha, Rodney NASA division chief for shuttle structural engineering at Johnson
Roe, Ralph NASA director of shuttle engineering at Johnson
Rominger, Kent chief astronaut at Johnson
Ross, Jerry veteran astronaut and NASA manager at Johnson

Schomburg, Calvin NASA engineer and shuttle heat tile expert at Johnson
Seriale-Grush, Joyce a NASA engineering manager at Johnson
Shack, Paul a NASA engineering manager at Johnson
Shriver, Loren deputy shuttle program manager for United Space Alliance and former astronaut
Simpson, Roger NASA liaison to the U.S. Strategic Command in Colorado
Smelser, Jerry NASA's external tank program manager at Marshall
Stich, Steve NASA flight director

Tetrault, Roger CAIB member and retired CEO at McDermott International
Turcotte, Stephen CAIB member and Navy rear admiral

Vaughan, Diane CAIB consultant and Boston College sociology professor

Wallace, Steve CAIB member and the FAA's director of accident investigation

Wetherbee, Jim veteran astronaut who helped oversee recovery of crew remains

Whittle, Dave chairman of NASA's Systems Safety Review Panel and Mishap Investigation Team

Widnall, Sheila CAIB member and engineering professor at the Massachusetts Institute of Technology

APPENDIX 2

LIST OF COMMONLY-USED ACRONYMS

CAIB	Columbia Accident Investigation Board
ET	external tank
FREESTAR	Fast Reaction Experiments Enabling Science, Technology, Applications and Research
FAA	Federal Aviation Administration
FEMA	Federal Emergency Management Agency
FRR	Flight Readiness Review
HQ	NASA Headquarters in Washington
IFA	In-flight anomaly
JSC	Johnson Space Center
KSC	Kennedy Space Center
LaRC	Langley Research Center
MADS	Modular Auxiliary Data System
MEIDEX	Mediterranean Israeli Dust Experiment
MER	Mission Evaluation Room
MMT	Mission Management Team
MOD	Mission Operations Directorate
MRT	Mishap Response Team
NASA	National Aeronautics and Space Administration
NIMA	National Imagery and Mapping Agency
ORR	Orbiter Rollout Review

PRCB	Program Requirements Change Board
RCC	reinforced carbon-carbon
SLA	super lightweight ablator
SRB	solid rocket booster
SSP	space shuttle program
STS	Space Transportation System
TPS	thermal protection system
UHF	ultra-high frequency
VAB	Vehicle Assembly Building at Kennedy Space Center
WLE	wing leading edge

APPENDIX 3

RECOMMENDATIONS OF THE COLUMBIA ACCIDENT INVESTIGATION BOARD

The Columbia Accident Investigation Board's final report was released Aug. 26, 2003. The board made 29 recommendations, 15 of which were to be implemented before shuttles could return to flight. Those items were designated "RTF."

Thermal Protection System

1 Initiate an aggressive program to eliminate all external tank thermal protection system debris-shedding at the source with particular emphasis on the region where the bipod struts attach to the external tank. [RTF]

2 Initiate a program designed to increase the orbiter's ability to sustain minor debris damage by measures such as improved impact-resistant reinforced carbon-carbon and acreage tiles. This program should determine the actual impact resistance of current materials and the effect of likely debris strikes. [RTF]

3 Develop and implement a comprehensive inspection plan to determine the structural integrity of all reinforced carbon-carbon system components. This inspection plan should take advantage of advanced non-destructive inspection technology. [RTF]

4 For missions to the international space station, develop a practicable capability to inspect and effect emergency repairs to the widest possible range of damage to the thermal protection system, including both tile and reinforced carbon-carbon, taking advantage of the additional capa-

bilities available when near to or docked at the international space station.

For non-station missions, develop a comprehensive autonomous (independent of station) inspection and repair capability to cover the widest possible range of damage scenarios.

Accomplish an on-orbit thermal protection system inspection, using appropriate assets and capabilities, early in all missions.

The ultimate objective should be a fully autonomous capability for all missions to address the possibility that an international space station mission fails to achieve the correct orbit, fails to dock successfully, or is damaged during or after undocking. [RTF]

5 To the extent possible, increase the orbiter's ability to successfully reenter Earth's atmosphere with minor leading edge structural sub-system damage.

6 In order to understand the true material characteristics of reinforced carbon-carbon components, develop a comprehensive database of flown reinforced carbon-carbon material characteristics by destructive testing and evaluation.

7 Improve the maintenance of launch pad structures to minimize the leaching of zinc primer onto reinforced carbon-carbon components.

8 Obtain sufficient spare reinforced carbon-carbon panel assemblies and associated support components to ensure that decisions on reinforced carbon-carbon maintenance are made on the basis of component specifications, free of external pressures relating to schedules, costs, or other considerations.

9 Develop, validate, and maintain physics-based computer models to evaluate thermal protection system damage from debris impacts. These tools should provide realistic and timely estimates of any impact damage from possible debris from any source that may ultimately impact the orbiter. Establish impact damage thresholds that trigger responsive corrective action, such as on-orbit inspection and repair, when indicated.

Imaging

10 Upgrade the imaging system to be capable of providing a minimum of three useful views of the space shuttle from liftoff to at least solid rocket booster separation, along any expected ascent azimuth. The operational status of these assets should be included in the launch commit criteria for future launches. Consider using ships or aircraft to provide additional views of the shuttle during ascent. [RTF]

11 Provide a capability to obtain and downlink high-resolution images of the external tank after it separates. [RTF]
12 Provide a capability to obtain and downlink high-resolution images of the underside of the orbiter wing leading edge and forward section of both wings' thermal protection system. [RTF]
13 Modify the memorandum of agreement with the National Imagery and Mapping Agency to make the imaging of each shuttle flight while on orbit a standard requirement. [RTF]

Orbiter Sensor Data

14 The Modular Auxiliary Data System instrumentation and sensor suite on each orbiter should be maintained and updated to include current sensor and data acquisition technologies.
15 The Modular Auxiliary Data System should be redesigned to include engineering performance and vehicle health information, and have the ability to be reconfigured during flight in order to allow certain data to be recorded, telemetered, or both as needs change.

Wiring

16 As part of the Shuttle Service Life Extension Program and potential 40-year service life, develop a state-of-the-art means to inspect all orbiter wiring, including that which is inaccessible.

Bolt Catchers

17 Test and qualify the flight hardware bolt catchers. [RTF]

Closeouts

18 Require that at least two employees attend all final closeouts and intertank area hand-spraying procedures. [RTF]

Micrometeoroid and Orbital Debris

19 Require the space shuttle to be operated with the same degree of safety for micrometeoroid and orbital debris as the degree of safety calculated for the international space station. Change the micrometeoroid and orbital debris safety criteria from guidelines to requirements.

Foreign Object Debris

20 Kennedy Space Center Quality Assurance and United Space Alliance must return to the straightforward, industry-standard definition of "for-

eign object debris" and eliminate any alternate or statistically deceptive definitions like "processing debris." [RTF]

Scheduling

21 Adopt and maintain a shuttle flight schedule that is consistent with available resources. Although schedule deadlines are an important management tool, those deadlines must be regularly evaluated to ensure that any additional risk incurred to meet the schedule is recognized, understood, and acceptable. [RTF]

Training

22 Implement an expanded training program in which the Mission Management Team faces potential crew and vehicle safety contingencies beyond launch and ascent. These contingencies should involve potential loss of shuttle or crew, contain numerous uncertainties and unknowns, and require the Mission Management Team to assemble and interact with support organizations across NASA/contractor lines and in various locations. [RTF]

Organization

23 Establish an independent Technical Engineering Authority that is responsible for technical requirements and all waivers to them, and will build a disciplined, systematic approach to identifying, analyzing, and controlling hazards throughout the life cycle of the shuttle system. The independent technical authority does the following as a minimum:

- Develop and maintain technical standards for all space shuttle program projects and elements
- Be the sole waiver-granting authority for all technical standards
- Conduct trend and risk analysis at the subsystem, system, and enterprise levels
- Own the failure mode, effects analysis and hazard reporting systems
- Conduct integrated hazard analysis
- Decide what is and is not an anomalous event
- Independently verify launch readiness
- Approve the provisions of the recertification program called for in recommendation [26]. The Technical Engineering Authority should be funded directly from NASA Headquarters, and should have no connection to or responsibility for schedule or program cost.

24 NASA Headquarters Office of Safety and Mission Assurance should have direct line authority over the entire space shuttle program safety organization and should be independently resourced.

25 Reorganize the Space Shuttle Integration Office to make it capable of integrating all elements of the space shuttle program, including the orbiter.

Organization

26 Prepare a detailed plan for defining, establishing, transitioning, and implementing an independent Technical Engineering Authority, independent safety program, and a reorganized Space Shuttle Integration Office as described in [23, 24 and 25]. In addition, NASA should submit annual reports to Congress, as part of the budget review process, on its implementation activities. [RTF]

Recertification

27 Prior to operating the shuttle beyond 2010, develop and conduct a vehicle recertification at the material, component, subsystem, and system levels. Recertification requirements should be included in the Service Life Extension Program.

Closeout Photos/Drawing System

28 Develop an interim program of closeout photographs for all critical subsystems that differ from engineering drawings. Digitize the close-out photograph system so that images are immediately available for on-orbit troubleshooting. [RTF]

29 Provide adequate resources for a long-term program to upgrade the shuttle engineering drawing system including:

- Reviewing drawings for accuracy
- Converting all drawings to a computer-aided drafting system
- Incorporating engineering changes

INDEX

Abbey, George, 219, 222
Air Force, Columbia space photo requests and, 110, 112–13, 118–19, 142–43
Al-Saud, Salman Abdulaziz, 28, 32
Altemus, Steve, 227, 229
Anderson, Michael P., 17, 45–47
  astronaut training of, 38–39, 44, 46
  as Columbia payload commander, 7, 8, 38, 45, 46, 47, 77, 78, 86, 89, 254, 256
Anderson, Sandra, 17, 46, 159
Apollo program, 27, 32–36, 45, 47, 55, 64, 67, 68, 73, 93, 119, 135, 165, 173, 246, 278–79, 287–88
Armstrong, Neil, 27, 287, 288
Ashkenazy, Haim, 30–31
Atlantis, 10, 33, 51, 236
  in Columbia crew survival scenario, 255, 256, 257–58, 259
Atlantis (STS-112), October 2002 launch of:
  electronics box hit in, 54–55, 59, 62
  film of, 53–54, 59, 62
  foam and debris strikes on, 59–63, 65–66, 68, 69, 71, 72, 76, 94, 95, 100, 101, 106–8, 144, 215
  IFAs and, 61–63, 65, 66, 206
Atlantis (STS-114), 176, 271, 272, 273, 277
  CAIB recommendations implementation and, 273–78
  commander and crew of, 271–73, 275–76, 277
  readying of, 106–7, 144–45
  safety of, 277, 278
Austin, Lambert, 63, 106, 109–10, 112, 113, 116, 211–12

Barksdale Air Force Base, 158, 168–69, 176–77, 181, 185, 198
Barry, John, 167, 177, 193, 213, 269
Beck, Kelly, 87–88
Benz, Frank, 11, 109, 116, 117, 118, 139, 211
Binder, Alan, 294–95
bipod ramps, 58–62, 66, 68–71, 72, 76, 85, 91, 94–95, 99, 100, 101–2, 106, 144, 215, 234, 235, 273, 276
Blevins, Gene, 16–17, 243
Boeing, 34, 97–98, 100–103, 121, 122, 126, 127, 128, 132, 206, 207
boundary layer, 238–39, 240, 241
Bradley, Theron, 167, 264
Bridges, Roy, 154–57, 200

Brown, Dave, 7, 17, 41, 43, 45, 48–50, 77, 78, 79, 85, 154, 157, 168
  astronaut training of, 38–39, 40, 50
  as Columbia mission specialist, 7, 8, 15, 38, 89, 254, 256
Bush, George W., 155, 160, 162, 163–64, 184, 209, 281
Bush, G. W., administration:
  CAIB probe and, 177, 210, 281
  civil space plans and, 209, 283–84, 292–93
Buzzard, Frank, 228, 259

Cabana, Bob, 25, 40, 42, 46, 78, 132, 153, 154, 158–59, 174, 186
CAIB (Columbia Accident Investigation Board), xiv, 189–285, 285
  Congress and, 177, 181, 190, 194, 281–285
  fact-finding trips of, 184–85, 193
  mission of, 76, 177, 181–82, 190, 191–92
  name of, 166–67
  and NASA management and safety culture, 195, 196, 199–206, 212–13, 214–15, 216, 217, 218–23, 266–67, 268–71
  NASA's assistance to, 167, 177, 186, 194, 210–11, 213, 228, 232–33, 264
  press conferences of, 186–87, 189–91, 255, 264, 267
CAIB, final report of, 177, 189, 192–93, 195, 203, 211, 214, 217, 223, 235, 261–85, 292
  findings and recommendations of, 265–68, 287, 289, 303–7
  NASA's broken safety culture implicated in, 265, 266–67, 268–70, 279–80
  Senate hearing on, 281–83
Cain, LeRoy, 3, 10–11, 13, 14, 19–23, 25, 119, 152–53, 232–33, 241, 248, 249, 254
Campbell, Carlisle, 125, 135–44, 215
Cape Canaveral Air Force Station, 78, 94, 100, 159, 161, 262
Card, Andrew, 155
Card, Mike, 142–43
CBS News, xii, 89–90
Challenger, 33, 58, 66
Challenger disaster (1986), xiii, 4, 9, 14, 26, 55, 60, 68, 76, 83, 96, 143, 152, 159, 165, 174, 268, 270, 278, 288
  cause of, 34, 67, 202
  Columbia parallels with, 202–3, 205, 215, 217–18, 223, 252
  crew recovery in, 174
  Reagan and, 33, 184, 281
  Rogers Commission and, 177, 178, 192, 194, 202, 219, 281
  wreckage of, 178, 262
Challenger Launch Decision, The (Vaughan), 202–3
Chandra X-ray Observatory, 36, 37
Chao, Dennis, 101
Chawla, Kalpana "KC," 17, 38–39, 43, 44, 45, 77, 78
  as Columbia flight engineer, 7, 15, 38, 43, 44–45, 82, 85, 199
Cheney, Dick, 186, 209
Clark, Iain, 17–18, 47, 74–75, 76, 84, 159
Clark, Jon, 17–18, 22, 47, 48, 74–75, 76, 84, 90, 153, 154, 158–59
Clark, Laurel, 38, 47–48, 90
  as astronaut, 38–39, 47, 48
  as Columbia mission specialist, 7, 12, 22, 38, 239
  countdown activities of, 77, 78
  family of, 17–18, 47, 48, 74–75, 153, 154, 158–59
  re-entry videotaped by, 9, 15, 199, 238, 239, 252
Clinton, Bill, 28
Clinton administration, 209
  space program and, 35, 222, 289
CNN (Cable News Network), 89–90
Coday, Frank, 172–73
Coday, Jeannie, 171, 172
Coday, Roger, 171–73, 254
Cold War, 287–88, 292
Cole, terrorist attack on, 160, 161, 191
Collins, Eileen, 271–73, 275–76, 277, 278

Columbia, 7, 8, 44–45, 89
age of, 10, 33, 187
technical problems and foam strikes on previous missions of, 10, 37, 51, 56, 58, 106, 207, 271
Columbia (STS-107), 36, 84–89, 98, 218
cracked fuel line bearings and, 75, 76
emergency landing scenarios for, 136–45
fueling of, 75, 77
heavier than normal landing weight of, 126, 129
MADS recorder of, *see* MADS recorder
media coverage inflight of, 89–90, 91, 129
recovery and storage of wreckage from, 12, 157, 162, 163, 169, 173, 178, 185, 186, 196–99, 225–30, 253, 262–63
Columbia (STS-107), crew of, 2–3, 27–32, 37–51, 74–81
"crew survivability" study and, 252–53
families of, *see* families of Columbia's crew
prayers and memorials for, 168, 173, 174–75, 184, 186, 187, 254
recovering remains of, 157–58, 172–74, 186, 198, 253–54
in shuttle's last minutes, 1–26, 241, 242, 244, 246–47, 248, 249, 251–54
survival scenarios for, *see* survival scenarios
Columbia (STS-107), foam strikes on, 73, 85, 90–91
as cause of Columbia disaster, 155, 165–66, 180, 183–84, 185–86, 211, 227, 233–37, 265–66, 267–68
debris assessment team analysis of, *see* debris assessment team
ground-based camera recordings of, 25, 91, 93–95, 96, 97, 100, 108, 136, 155, 165, 233–34, 274–75
landing and reentry concerns and, 98, 100, 101–2, 103, 106, 109, 120, 132–33, 135–45, 180
location and severity of, 94–95, 96, 97, 98–99, 100–103, 105, 106, 108, 116, 120, 121–22, 127, 129–31, 183, 227–28, 230–33, 234, 236, 255, 265–66, 274
management minimizing of, 20–21, 90, 98–99, 106, 115, 119, 121, 127, 128, 131–32, 133, 134–35, 138, 143, 147, 165–66, 171, 180–81, 185–86, 212–13, 255
origin of, 94, 99, 100, 234
as potential threat to future missions, 105–6, 108, 115, 130, 218
RCC damaged by, *see* RCC damage (on Columbia STS-107)
requests for space photos of, 95–96, 102, 103, 109–19, 120–21, 122, 128, 134–35, 136, 137, 142–43, 166, 180–81, 183, 200, 204–5, 213, 214–15, 254, 255
size and shape of, 94, 95, 96, 100–101, 102, 103, 105, 108, 116, 121, 127, 129, 183, 211, 233–34, 237, 240
Columbia (STS-107), launch of, 73–91
delays in, 37, 38, 50–51, 86
environmental conditions and, 76, 77, 235
liftoff of, 81–82
MADS and, 234
Mission Control and, 80–83
post-9/11 security precautions and, 74, 75, 78, 79
roll maneuvers and, 82, 84
and trip to launchpad, 73–74, 78–79
Columbia (STS-107), re-entry and break-up of, 1–26, 147–69, 237–53, 272, 288
astronaut videotaping of, 9, 197, 199, 238, 239
atmospheric friction and heating in, 1, 2, 6, 8–9, 12–13, 16–17, 98, 149, 150, 230, 231, 238, 239–51
automated descent in, 7, 8, 11, 15, 17, 20, 21, 179–80, 183, 233, 240, 242, 244, 245, 246, 251
causes of, 213, 265–68, 292

Columbia (STS-107), re-entry and break-up of (*cont.*)
Challenger disaster parallels with, 202–3, 205, 215, 217–18, 223, 252
communications dropout and loss in, xxi, xii–xiii, 22, 23, 24, 25, 26, 150, 151, 152, 165, 180, 233, 247, 248, 249
crew module in, 251–53, 263
eyewitnesses and videos of, 16–17, 147–50, 151–52, 180, 186, 227, 233, 243, 244, 245, 246, 249
falling debris and footprints in, 148, 149–50, 157, 162, 163, 225–30, 243, 244, 245, 246, 249, 253
investigation into, *see* CAIB
landing gear sensors lost in, 23, 150, 165, 179, 196, 227, 242, 245, 247
left wing failure in, 239, 240–44, 246–47, 249, 250, 255
left wing sensors lost in, xii, 19–22, 23, 149, 150, 155, 164, 165, 179–80, 182, 234, 239, 240–41, 243, 245
media coverage of, xi–xiii, 16–17, 18–19, 26, 154, 161–62, 164–66, 173, 175, 177–81, 182–84, 185–86, 187
Mission Control and, xii–xiii, 3, 5–6, 10–13, 14, 16–17, 19–24, 25–26, 140, 152–53, 158, 178–80, 182–84, 233, 241, 242, 244, 245, 247–49
NASA's television coverage of, xi, 6, 23–24, 148, 152, 154, 155, 156, 164
path of, xi, xiii, 2, 6, 8, 12, 13, 15, 16, 17, 20, 21, 23–24, 147–50, 157, 158, 179, 227, 233, 241, 242, 243–44, 245, 246
runway for, xi, 5–6, 10, 13–14, 17, 21, 24, 25–26, 141, 153, 240, 245
sonic boom in, 17, 148, 150, 172
tire pressure loss indicated in, xii, 22–23, 150–51, 165, 179, 245–46, 247, 248
velocity of, 1–2, 6, 9, 10, 12, 16, 17, 21, 149, 165, 176, 228, 234, 242, 245, 251

Congress, U. S., 86
CAIB and, 177, 181, 190, 194, 281–85
and NASA accountability issue, 212, 217, 279, 281–83
and NASA funding, 33, 284, 285, 290, 292, 293, 296
Conte, Barbara, 118–19
Contingency Action Plan, 152–69
CAIB and, *see* CAIB
Mishap Response Team and, *see* Mishap Response Team
purpose of, 26, 155
Covey, Dick, 279–80, 285
Crater (computer model), 101–2, 108, 121, 122, 127, 129, 183, 207
Crippen, Bob, 187
cryopumping, 234–35
Curry, Don, 120, 126

Daugherty, Bob, 136–42
Deal, Duane, 167, 193, 269–70
debris assessment team, 99–123, 233–34
Columbia space photo request made by, 103, 109, 113–15, 116–19, 120–21, 122, 128, 200, 201, 204–5
conclusions of, 121–22, 123, 128, 133, 134, 139, 143, 180, 181, 235, 255
damage scenarios of, 121–22, 127–30, 131–36, 142, 183
at MER presentation, 120, 125–28
Diggins, Dan, 204–5, 214
Discovery, 10, 28, 33, 51, 65, 95, 236
Dittemore, Ron, 11, 63–66, 76, 96, 105, 113, 118, 129, 156, 171, 194, 200, 208, 211, 217, 219–20, 235
Columbia space photos requests and, 109, 111, 134–35, 204, 205
Endeavour's FRR (2002) and, 65, 69–70, 71, 72, 106–7
news briefings of, 164–66, 178–81, 182–84, 185–87, 189
Dover Air Force Base, 174, 186
Drewry, Doug, 118, 134
Dunham, Mike, 122, 126, 127, 132–33

Edwards Air Force Base, 4, 13, 120
EI (entry interface), 6, 8–9, 238–39, 240

elevons, in Columbia's final minutes, 19, 242, 244, 245, 246, 247, 248, 249
Endeavour, 10, 51
November 2002 launch of, 53, 62, 63, 65, 66, 67–71, 72, 106–8, 113, 157, 215
Engelhauf, Phil, 11, 90, 110, 112, 119, 129, 152
EPA (Environmental Protection Agency), 163, 174
European Space Agency, 35, 85, 87
explosive bolts, 62, 67, 81–82

FAA (Federal Aviation Administration), 167, 177, 204
families of Columbia's crew, 158–60, 161, 162, 184
at launch, 75–76, 81, 82, 84
remains released to, 174
at shuttle arrival site, 6, 17–18, 22, 26, 153–54, 155
FBI (Federal Bureau of Investigation), 103
FEMA (Federal Emergency Management Agency), 163, 174, 228
Feynman, Richard, 202
flight controllers, 138, 139–41
foam and debris strikes, 57, 69, 70, 115
on Atlantis, 59–63, 65–66, 68, 69, 71, 72, 76, 94, 95, 100, 101, 106–8, 144, 215
from bipod ramps, *see* bipod ramps
as common on shuttle launches, 55–56, 58, 59, 61–62, 71, 259
hazard report on, 57, 70–71, 115, 200
to heat-shield tiles, *see* heat-shield tiles
as IFAs, 61–63, 65, 66, 206
as "out of family," 57–58, 62, 99
to RCC surfaces, *see* RCC surfaces, foam and debris strikes on
see also Columbia (STS-107), foam strikes on
forward spar, 231, 239, 240–44, 245
FREESTAR, 88
FRRs (Flight Readiness Reviews), 53, 65, 66, 67–71, 72, 75, 106–8, 215

Garske, Ed, 150, 151–52
Gehman, Harold W., Jr., 160, 190–91
CAIB and, 160–61, 166–67, 168, 175, 176–77, 181–82, 184–85, 190–92, 193, 194, 195–96, 200–201, 206, 207, 209, 210–23, 254–58, 263–64, 265, 267–68, 279, 281–85, 289
Gemini program, 64, 135
Goldin, Dan, 37, 219, 221–23, 293
Goodman, Pat, 241, 243, 245, 246, 247
Gore, Al, 36–37, 222
Gregory, Fred, 160–61, 166, 174, 177, 215–17, 219, 222, 262

Hale, Wayne, Jr., 18, 93, 94–97, 109–14, 131, 154–57, 211
Hallock, Jim N., 167, 193, 194, 238, 244–45, 246–47, 285
Ham, Linda, 11, 13, 20–21, 76–77, 96, 103–8, 283
CAIB investigation and, 210–11, 213, 214–15
Columbia space photos request and, 110–12, 114, 115, 204, 205, 214, 282
Mishap Response Team and, 156, 210–11, 213
MMT and, 103–8, 128–32, 133–34, 166, 210, 211, 212–13, 214–15
Har-Even, Aby, 28, 29
Harrison, Jean-Pierre, 17
Hartenstein, Bill, 16–17
Hartsfield, James, 3, 6, 11, 13, 24, 26, 148, 152, 153
heat-shield tiles, 12, 15, 56, 238
damaged on Columbia's previous missions, 56, 58, 106
damaged on Columbia STS-107, 101–2, 108, 116, 119, 120, 121, 122, 127, 129–31, 180, 183, 229, 230, 240, 243, 247
foam damage to, 55–56, 57, 58, 59, 71, 77, 97, 98, 101, 106, 119, 228
impact tests on, 211, 235–37
orbital repair of, 111, 120, 135, 165–66, 180, 274
purpose of, 12, 55, 56, 98, 130, 230
zipper effect and, 132

Heflin, Milt, 164–65, 203
Hess, Ken, 167, 193, 285
Higgins, Bill, 200–201
Hill, Paul, 186, 225, 233
Hobaugh, Charles, 22–23, 24, 26, 82, 248
Hollings, Ernest, 282–83
Holloway, Tommy, 64–65
Homeland Security, 155, 162
Hubbard, Scott, 167, 192, 193, 194, 211, 236–37, 250, 268, 279
Hubble Space Telescope, 34, 36, 274, 293
Husband, Evelyn Neely, 4, 17, 46, 153–54, 158
Husband, Rick, 2–5, 74, 77, 78, 79
  as Columbia's commander, 2, 3, 6–9, 10, 12, 14, 15, 16, 18–19, 22, 23, 24, 38–39, 43, 44–45, 80, 82, 83, 84, 86, 89, 90–91, 137, 140, 151, 180, 183, 199, 241, 242, 244, 246, 248, 249, 251, 252, 272
  family of, 2, 4, 10, 17, 40, 47, 158
  religion and spirituality of, 2–3, 5, 7, 10, 41, 46, 175, 272

ice, as shuttle tank hazard, 58, 77, 273
IFAs (in-flight anomalies), 9
  Atlantis 2002 launch and, 61–63, 65, 66, 206
Inconel, 231, 240
Integrated Hazard Report 37, 57, 70–71, 115, 200
Intercenter Photo Working Group, 24–25, 55, 59, 61, 68, 94, 95, 97, 100, 103, 205
International Space Station, 32–36
  budget schedule pressures on, 105–6, 218–21, 222, 266
  costs of, 35, 209, 218, 221, 222, 290, 291–92, 296
  as emergency safe haven, 276–77
  partners in, 35, 218, 263, 273, 291, 292, 296
  shuttles and, 51, 60, 71, 105–6, 107, 176, 177, 218, 222, 254, 255, 263, 271, 272–73, 274, 275, 296
International Space Station and Shuttle Mishap Interagency Investigations Board, *see* CAIB
Iraq, 8, 30, 210
Israel, 8, 30–31, 74, 75
  see also Ramon, Ilan
Israel Space Agency, 28–30, 37–38
Issacharoff, Dean, 27–28, 29
Issacharoff, Jeremy, 27–28

Johnson Space Center, 4, 5–6, 10–13, 14, 16–17, 19–24, 25–26, 31, 41, 54, 64, 95, 96, 97, 99–100, 103, 108, 139–41, 175, 184, 190–91, 220, 232–33
  Action Center of, 103, 126, 128
  management shakeup at, 211–12, 217, 281, 282–83
  Mechanical Systems Group at, 139
  MER (Mission Evaluation Room) at, 97–99, 103, 105, 110, 111, 114, 120, 125, 143, 150–53

Kelly, Jim, 273, 277–78
Kennedy Space Center, 2, 220
  Columbia reconstruction at, 227, 229, 231–32, 239, 262
  as Columbia's re-entry destination, xi, 3, 5–6, 8, 10, 12, 13–15, 17–19, 20, 21, 23, 24, 25, 120, 141, 240, 245
  crew quarters at, 77, 153, 158–59, 160
  Image Analysis Facility of, 53–54, 59, 68, 94, 95, 97, 100
  Launch Control Center at, 78, 81, 82, 84
  Operations and Checkout Building of, 67, 77–78, 153, 158–59
  post-9/11 security at, 74, 75, 78, 79
  public affairs center of, xi–xii, 18, 23, 161
  Shuttle Landing Facility of, 6, 14–15, 153, 154, 187
  Vehicle Assembly Building of, 54, 73, 107, 144, 262–63
King, Dave, 162–63
Kirtland Air Force Base, 148–49

Kling, Jeff, 11, 19, 21–22, 23, 167–68, 247–48
Kostelnik, Mike, 218–20, 283
Kowal, John, 135, 136
Kranz, Gene, 93, 96, 278–79

Law, Howard, 137, 138
Lawson, Jay, 147–48
Lechner, David, 139–40
Leinbach, Mike, 80, 154–57, 158, 168–69, 229
Lockheed Martin, 34, 57, 97–98, 206, 215, 290
Logsdon, John, 192, 194, 202, 222–23
Lu, Ed, 272–73

McAuliffe, Christa, 14, 34
McCain, John, 282
McCluney, Kevin, 140–41
McCool, Lani, 17, 39, 42, 153, 160, 189
McCool, William "Willie," 41–42, 77, 78, 81
  astronaut training of, 38–39, 40
  as Columbia pilot, 7, 8–9, 12, 15, 16, 19, 38, 82, 83, 84, 89, 189, 199, 242, 244, 251
  family of, 8, 17, 39, 42, 47, 153, 189
McCormack, Don, 98–99, 105–6, 107, 150–52, 126–31, 133–34
McCulley, Mike, 208
Madera, Pam, 99, 102, 103, 126, 132, 135
MADS (modular auxiliary data system) recorder, 225–28, 232, 233, 234, 263, 270
  in Columbia's final minutes, 239, 241, 242, 248–49, 251
Malenchenko, Yuri, 272–73
Mars, exploration of, 167, 221, 288, 291, 294, 295–96
Marshall Space Flight Center, 54, 55, 66, 93, 97, 162, 220
  External Tank Project at, 57, 60–61, 67–68, 71–72, 184, 206, 211, 215
MEIDEX (Mediterranean Israeli Dust Experiment), 28–29, 32, 37, 88
Mercury program, 64, 173
Merritt Island tracking station, 24, 25, 152
Michoud Assembly Facility, 57, 58, 61, 185, 215
Micklos, Ann, 15, 41, 48, 49, 79, 154, 157
Mir space station, 35
Mishap Response Team, 156–58, 162, 168–69, 173–74, 228
  crew remains recovered by, 157–58, 172–74, 185, 186, 198, 253–54
  debris recovery and, 157, 162, 163, 169, 173, 178, 185, 186, 196–99, 225–30, 253
MMT (Mission Management Team), 11, 13, 20–21, 65, 76–77, 96–97, 103–8, 116, 143
  CAIB as critical of, 203–5, 212–13, 215, 267, 268
  debris assessment team and, 105–6, 121, 122, 126, 128–32, 133–34, 204–5, 212, 214–15
Moon, Darwin, 97, 101
moon, exploration of, 27, 47, 246, 287, 288, 294–95
Morgan, Barbara, 14, 159

NAS (National Academy of Sciences), 37
NASA (National Aeronautic and Space Administration):
  Ames Research Center of, 137, 141, 207
  budget and finances of, 33, 35, 101, 206, 209–10, 221, 222–23, 279, 283–85, 288, 290, 291–92, 293, 295
  CAIB and, *see* CAIB
  central mission of, 292–93, 296
  contractors and, 206–8
  disaster contingency plan of, *see* Contingency Action Plan
  "Goldin years" at, 222–23
  Israel Space Agency agreement with, 28–29, 37–38
  Langley Research Center of, 136, 138, 139, 144, 212
  poor management and safety culture of, 199–223, 266–67, 268–71, 279–83

NASA (National Aeronautic and Space Administration) (*cont.*)
post-Columbia management shuffling at, 211–12, 217–18, 281, 282–83
post-9/11 security of, 74, 75, 78, 79
Safety and Mission Assurance Office of, *see* Safety and Mission Assurance Office
space stations and, 33–35, 209, 288
see also Johnson Space Center; Kennedy Space Center; Marshall Space Flight Center
NASA Television, xi, 6, 23–24, 148, 152, 155, 156, 164
National Air and Space Museum, 27, 29
National Imagery and Mapping Agency, 110, 142, 233–34
Nixon, Richard, 33
Noguchi, Soichi, 273, 277
Norman, Ignacio, 128, 132, 135

Ochoa, Ellen, 11, 152
O'Connor, Bryan, 70, 108, 142–44, 160, 215, 219, 283
CAIB and, 167, 177, 194, 264
O'Keefe, Sean, 25–26, 154–57, 160–63, 168, 186, 191, 209–10, 215–16, 217, 218, 219, 220, 221, 222, 228, 264, 293
CAIB and, 209–11, 213–14, 264, 265, 279–83
Oliu, Armando, 54–55, 62, 77, 94, 95, 100
Orbiter Recovery Convoy, 14–15, 153, 154
O-ring seals, 67, 202
Orlando Sentinel, xii, 183, 195, 281, 294
Ortiz, Carlos, 97, 100–101, 102–3, 108, 126–27
Osheroff, Doug, 192, 193, 194, 235
Otte, Neil, 60–61, 62–63, 66, 67

Page, Bob, 24–25, 55, 59, 61–62, 66, 77, 94, 103, 131
Columbia space photo request of, 95–96, 110, 112, 113, 115, 204–5
Palestine, Tex., 196–97, 198
Parker, Paul, 101–2, 103
Pastorek, Paul, 154–57
Patrick Air Force Base, 95, 110
Peterson Air Force Base, 112
Petete, Trish, 113–14, 203, 212
Phillips, Dave, 110, 111, 112
Pitre, Dave, 5, 45
Pontecorvo, Marty, 196–97, 199
pressure suits, 7, 15, 77–78, 253
Program Requirements Control Board, 61–66, 206

Racz, Darla, 27, 39, 50
Ramon, Ilan, 7, 8, 18, 29–32, 35–40, 74, 77, 78, 79, 89
as Columbia payload specialist, 7, 8, 9, 28–29, 32, 36, 37–38, 88
family of, 17, 30, 31, 47, 75, 84, 159
Ramon, Rona, 17, 31, 159
RCC damage (on Columbia STS-107), 91
debris assessment team analysis of, 102, 121, 122, 128, 130, 131, 233, 235
emergency repair scenarios for, 254, 255, 256–57
in final minutes, 238–42, 245
impact tests and, 211, 236–37
MADS and, 234, 239
recovered wreckage showing, 229, 231–33, 263
RCC (reinforced carbon-carbon) surfaces, 12, 56, 98, 231–32
CAIB recommendations for orbital repair of, 274, 276
foam and debris strikes on, 56, 59, 98, 101, 122, 235, 236–37
resiliency of, 98, 101, 128
Readdy, Bill, 23, 25, 26, 139, 142–43, 53, 67, 70–71, 144, 168, 215, 219
contingency plan and, 154–57, 161, 168
Reagan, Ronald, 33, 34, 35, 184, 281
Reed, Lisa, 41, 43, 49–50
Return to Flight Task Group, 279–80, 285
Ride, Sally, 192, 193, 194, 195, 196
Ridge, Tom, 155, 162
Robinson, Steve, 273, 277

robot arms, 165, 274, 275, 276
robotic probes, 293–94, 295
Rocha, Robyn, 100, 151
Rocha, Rodney, 99–100, 107–9, 113–15, 116–20, 122–23, 126, 128, 131–33, 136, 139, 143, 144, 151, 201, 204, 247, 267
rockets, unmanned, U.S., 33, 34
Roe, Ralph, 11, 20–21, 118, 119, 210–11, 213, 236, 283
  Columbia space photos request and, 109, 110–11, 113, 114–15, 117, 282
Rominger, Kent, 3, 14, 42, 44–45, 78, 258
Ross, Jerry, 25, 132, 153, 154, 158, 159–60, 161, 199, 207, 222, 258–59
Russia, 32, 35, 176, 263, 291

Safety and Mission Assurance Office, 70, 139, 142–43, 215–17, 219, 269, 281, 282, 283
Saudi Arabia, first astronaut from, 28, 32
Schomburg, Calvin, 98, 110–11, 114, 119–20, 121, 126, 130, 131, 132, 133, 134–35, 144, 236
Scientific Applications International Corporation, 102
Senate, U.S., 200, 281–85
Seriale-Grush, Joyce, 99–100, 117, 119, 126, 133, 139, 151
Shack, Paul, 102, 107–8, 113–14, 116–17, 119, 126, 133, 139, 214
Shannon, John, 11, 255–56, 257
Shriver, Loren, 111, 126, 204
Shuart, Mark, 138, 139
Shuttle Training Aircraft, 3, 14, 78
Simpson, Roger, Columbia space photos request and, 112–13
Skylab space station, 32, 33, 288
SLA (super lightweight ablator), 60–61
Smelser, Jerry, 67–70, 71, 106–8, 215
Southwest Research Institute, 103, 236–37, 268
space exploration, U.S. future in, xiii, 292–97
Space Flight Operations Contract, 206
Spacehab module, 12, 35–36, 86–89, 90
Spacelab, 35
space shuttles, 2, 9, 33–36, 137
  CAIB recommendations for, 287, 289
  costs of, 33, 206, 222, 283–85, 291, 296
  Critical Items List and, 205–6, 213
  declines in budget, workforce, and safety of, 101, 206–8, 213, 222–23, 269, 279, 283, 284–85, 292
  escape systems for, 253, 270
  foam and debris strikes on, *see* foam and debris strikes
  grounding of, 4, 34, 51, 55, 60, 65, 67, 208, 263, 272–73
  and IFA designation, 61–63, 65, 66, 206
  recertification of, 289
  reentries and landings of, 9–10, 13–14, 18, 22, 23, 46–47, 141, 238–39, 287–88
  replacements for, 289–90, 296
  safety and reliability of, 9, 277–78, 288–89
  space stations and, 32–36, 51, 60, 71, 105–6, 107, 176, 177, 218, 222, 254, 255, 263, 271, 272–73, 274, 275, 296
  speed of, 1–2, 6, 9, 10
  thermal protection system of, *see* heat-shield tiles; RCC
  upgraded camera systems for, 274–276
  wing design of, 230–31
space stations, manned U.S., 32–35
  see also International Space Station
spacewalks, 33, 165, 254, 255, 256–57, 272, 273, 277, 293
Stafford, Tom, 279–80, 285
Stich, Steve, 73, 90–91, 112
Stoner, Mike, 97–98
STS (Space Transportation System), 33
survival scenarios, 254–59

telescopes, military, 95, 103, 137, 180, 183
terrorism:
  Cole bombing, 160, 161, 191
  Columbia disaster and, 155
  September 11 attacks, 74

Tetrault, Roger, 192, 194
Texas, eastern, Columbia recovery operation in, xiii, 149–50, 157, 158, 171–74, 196–99, 225–27, 228, 229, 233, 247, 253–54
Texas Forest Service, 198–99
Toledo Bend Reservoir, 150, 198
Triana satellite, 36–37
Turcotte, Steve, 167, 193
Turner, Stephanie, 40, 41

United Space Alliance:
- CAIB criticism of, 206–7
- Columbia space photos request of, 109, 110, 111, 113, 115, 118
- debris assessment team and, 97–98, 100–103, 118, 126

U. S. Strategic Command, 110, 112–13

Vaughan, Diane, 202–3, 215, 217–18, 281
Vita, Carl, 196–97, 199

Wallace, Steve, 167, 176–77, 193, 207, 214, 270
Wetherbee, Jim, 157, 162, 173–74
Wetmore, Mike, 26, 154–57
White, Bob, 109, 110
White, Doug, 126, 246
Whittle, Dave, 156, 158, 169, 186, 226, 227
Widnall, Sheila, 192, 193, 194, 265, 285

X-33, 290

Young, John, 11, 246